Common Algorithms in Pascal

with Programs for Reading

Prentice-Hall Software Series
Brian W. Kernighan, advisor

Common Algorithms in Pascal

with Programs for Reading

DAVID V. MOFFAT

PRENTICE-HALL, INC., Englewood Cliffs, New Jersey 07632

Library of Congress Cataloging in Publication Data

Moffat, David V.
Common algorithms in Pascal with programs for reading.

Bibliography: p. 225
Includes index.
1. PASCAL (Computer program language) 2. Algorithms.
I. Title.
QA76.73.P2M63 1984 001.64'25 82-18623
ISBN 0-13-152637-5

Production supervision by Linda Mihatov
Manufacturing buyer: Gordon Osbourne

TO SUSIE AND MEGHAN

Printed in the United States of America

10 9 8 7 6 5 4 3 2 1

ISBN 0-13-152637-5

Prentice-Hall International, Inc., *London*
Prentice-Hall of Australia Pty. Limited, *Sydney*
Editora Prentice-Hall do Brasil, Ltda., *Rio de Janeiro*
Prentice-Hall Canada Inc., *Toronto*
Prentice-Hall of India Private Limited, *New Delhi*
Prentice-Hall of Japan, Inc., *Tokyo*
Prentice-Hall of Southeast Asia Pte. Ltd., *Singapore*
Whitehall Books Limited, *Wellington, New Zealand*

Contents

Preface

PURPOSE

This book is a collection of algorithms that are common to a wide variety of computer programming applications. It gives the reader a vocabulary or repertoire of useful algorithms, bringing together and identifying more algorithms of general utility than any programming text. It is intended to be used as a reference or as a supplement to a programming text. It smooths the transition from introductory programming to the formal study of algorithms.

The book also serves as a source of examples, exercises, and readings. Many exercises ask the reader to think about the algorithms or to write variations. Each part of the book includes one or more completely self-contained ''programs for reading'' that show practical applications of many of the algorithms. These are followed by more exercises that offer motivation for reading the programs.

WHO CAN USE THIS BOOK

Any person who is learning to program in Pascal will want eventually to know these common algorithms. In particular, students in first and second programming courses should find it a useful supplement, whether it is assigned reading or not; it contains most or all of the algorithms taught in those courses.

Instructors of these or of other courses will find here a good source of blackboard examples, reading exercises, written exercises, and test questions.

Programmers who use other languages, but who wish to learn Pascal, can see how the common algorithms are expressed in Pascal.

ORGANIZATION

The material is presented in an order common to many introductory texts so that it will complement rather than conflict with them. (The introduction to Part I lists the first occurrence of each language feature.) In particular, the book relegates procedures and functions to the latter half because many texts do so. Some very basic material is included for beginning programmers. In each section I have arranged the algorithms so that the easier ones appear first. In some sections it was also possible to show how complex algorithms may be constructed from simpler ones.

Part II contains the only major departure from the ''typical'' ordering of material: the second half presents simple state-transition techniques (without lookahead) for solving character-processing problems. This subject is too useful to be omitted, but it can also be left for later reading.

Part VIII presents some information-hiding techniques and other special topics.

The algorithms are indexed and cross-indexed.

PROGRAMMING STYLE

The algorithms are written with ''header'' comments identifying their purposes and with step-by-step comments explaining how they work. The reader is expected to understand the algorithms by reading them together with these comments, rather than by reading a separate prose paragraph. Drawings illustrate the more difficult ones.

The style of layout, indention, and commenting are entirely consistent throughout the book, so that the reader can become familiar with it. The stylistic details were carefully selected to reflect the program structure and to make a clear distinction between the descriptive material and the algorithms.

This is one of the very few books presenting programs that stand by themselves. Both the internal and the more often external documentation are included *within* each example program.

THE LANGUAGE USED

All the programs were compiled and executed by the OS version of the Waterloo Pascal interpreter running under MVS on an IBM 3081 at the Triangle Universities Computation Center, Research Triangle Park, North Carolina.

Every attempt was made to present only standard Pascal constructs as specified by the International Standards Organization Draft Proposal 7185, which supersedes the original *Pascal User Manual and Report* by Jensen and Wirth (see the Bibliography), although the two are quite similar. In particular, all the examples of input and output assume a batch environment. None of the algorithms rely upon the nonstandard aspects of the Waterloo implementation.

ACKNOWLEDGEMENTS

I would like to thank the many reviewers who improved the manuscript. I especially appreciate Lionel Deimel's comments and enthusiasm.

Don Martin, our department head, supported the project from its inception and provided the means for testing the book in our classrooms at North Carolina State University, Raleigh.

Many of the programs for reading were prepared by John Potok. He also provided answers to the exercises.

The text was phototypeset on a Compugraphic EditWriter 7700 at the Printing and Duplicating Department of the University of North Carolina, Chapel Hill. The copy was prepared with the SCRIPT text formatting program, a product of the University of Waterloo, Ontario. The SCRIPT to Compugraphic translation was arranged by the Computation Center of the University of North Carolina, Chapel Hill. The prose is set in Baskerville, the code in OCR-B.

David V. Moffat

Part I: General Algorithms

INTRODUCTION AND OVERVIEW

An algorithm is a finite sequence of well-defined actions whose result is to accomplish a given task. If they are to be put to any use, whether as subjects of study or as instructions to a computer, algorithms must be expressed in a spoken or written language or notation. Whenever an algorithm is written in a programming language, it bears the imprint of that language. A page-long algorithm in one language, for example, may be a single statement in another—a "sequence" of one action! Nonetheless, an algorithm has some kind of expression in any language.

A large set of algorithms are common to a broad range of computer applications and are expressed in a variety of programming languages. Taken together, they form a kind of language themselves, in which the solutions to many programming problems can be described. This "language" of common algorithms, not the knowledge of a particular programming language, is the key to success in programming. This book expresses that language of algorithms in Pascal.

Very few problems that can be solved with computers are solved merely by stringing together some algorithms; there is more to programming than that. The solution to a problem is described in terms of some central algorithms. These are usually supported by other algorithms, such as to input data and to display results. Many problem solutions also refer to an environment or situation that must be simulated within the program. Finally, the objects and terminology of the problem area and its solution must be defined as constants, types, and variables in the program.

Yet algorithms are still the basic building blocks of programs, and familiarity with the common algorithms is basic to the act and art of programming.

You can use this book in two ways: as a reference or as an introduction to algorithms in conjunction with an introductory text. In either case, it is best to read all of Part I to become familiar with the method of presentation and with some basic terms and algorithms that are echoed throughout the book.

If you use the book as a supplement to a text, then try to read it in the same order as the material in your text. However, if you chance upon some unfamiliar material here, you can be sure that you will eventually want to learn it anyway; there are no "cute" or "one-shot" illustrative examples, nor any esoteric applications. All the algorithms are used again and again in general programming. You can also read each part in its entirety to find out "all about" the algorithms that apply to the given data structure.

If you use the book as a reference, you will find that each algorithm is indexed, cross-indexed, and identified with comments. Each part is as self-contained as possible, given the fact that many algorithms have variants for several data structures. It is important to recognize that, although some algorithms can be copied unchanged into any program, many algorithms serve only as outlines or skeletons that must be filled in with details specific to each program.

The outline that follows shows where each of t
introduced in this book. It can be used to determ
want to know to read each part:

Part I: VAR, CONST, INTEGER, RE
BEGIN-END, READ, READLN, WRITE,
and simple expressions. The programs for re
CASE in simple contexts.

Part II: Set constants using character subrang
TRUE, FALSE, TYPE for defining enumera
one procedure. The example programs use RE

Part III: General TYPE definitions, integer su
FOR with DOWNTO, and one RECORD in the last section.

Part IV: Matrices, nested FOR loops, RECORD, ARRAY of RECORD, WITH, and one PROCEDURE. The program for reading contains several PROCEDUREs.

Part V: Strings, a PROCEDURE or a FUNCTION for each algorithm, and the INPUT buffer variable in a simple application.

Part VI: FILE, file buffer variables, GET, PUT, REWRITE, RESET, and the full syntax for READ and WRITE.

Part VII: Pointers, NEW and DISPOSE, and NIL.

Finally, try to do the exercises—or at least think about them—as you encounter them; they are designed and strategically placed to reinforce your understanding of the algorithms. Answers and hints for exercises that are marked with "†" are given in the back of the book.

SIMPLE COMPARISONS AND EXCHANGES

As you will see throughout this book, comparisons and exchanges of values are important classes of algorithms. In this section we start with the simplest ones.

Assume these declarations:

```
VAR
    A : INTEGER;              (* A, B, and C are any arbitrary  *)
    B : INTEGER;              (*    integers.                   *)
    C : INTEGER;
    SMALL : INTEGER;          (* SMALL and HOLD will be         *)
    HOLD : INTEGER;           (*    explained in the discussion.*)
```

and assume that A, B, and C have been assigned values.

The first algorithm selects one of two values:

```
(* Save the smaller of two values in SMALL:                    *)
IF A < B THEN
   SMALL := A
ELSE
   SMALL := B
```

Often, given one small value, we will want to find a still smaller value:

```
(* Save the value of A if it is smaller than SMALL:            *)
IF A < SMALL THEN
   SMALL := A
```

So we could then have:

```
(* Assign to SMALL the smallest of three values:               *)
IF A < B THEN
   SMALL := A
ELSE
   SMALL := B;
IF C < SMALL THEN
   SMALL := C
```

Ex. 1: Tell why this does not make sense:

```
IF A = SMALL THEN
   SMALL := A
```

and why this might make sense:

```
IF A <= SMALL THEN
   SMALL := A
```

There is a simple-looking algorithm called a *shift* that has important applications. Assume that A and B have been the variables of interest and that C contains the "next" value of interest:

```
(* Shift the value of C into the A B pair, losing A:        *)
A := B;
B := C
```

The shift would be diagrammed this way:

A ← B ← C

Its applications will be seen shortly.

All the preceding algorithms *substitute* a new value for the current value of a variable. The old value is lost. We often must instead *exchange* the values of two variables:

```
(* Exchange (or "swap") the values of A and B:              *)
HOLD := A;
A := B;
B := HOLD
```

where HOLD is any variable set aside for this purpose.

†*Ex. 2: How many extra variables like HOLD would you need to exchange the values of three variables? Try it this way:*

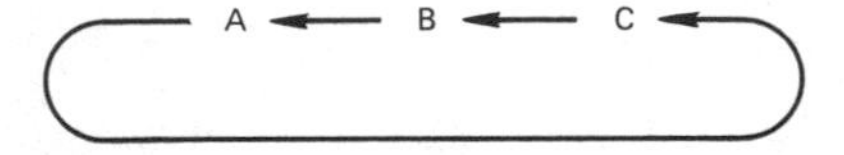

Exchanges are always done with some purpose in mind. For example:

```
(* Arrange the values of A and B so that A is smaller:     *)
IF A > B THEN
   BEGIN (* exchange *)
   HOLD := A;
   A := B;
   B := HOLD
   END (* exchange *)
```

Ex. 3: Does this algorithm guarantee that A is less than B?

This rearrangement of values is called *ordering* or *sorting.* We can order or sort any number of values:

```
(* Order the values of A, B, and C from smallest (in A)   *)
(* to largest (in C):                                      *)
IF B < A THEN
   BEGIN (* swap A & B *)
   HOLD := A;
   A := B;
   B := HOLD
   END; (* swap *)
(* Now A <= B.                                             *)
IF C < A THEN
   BEGIN (* swap A & C *)
   HOLD := A;
   A := C;
   C := HOLD
   END; (* swap *)
(* Now A <= B and A <= C.                                  *)
IF C < B THEN
   BEGIN (* swap B & C *)
   HOLD := B;
   B := C;
   C := HOLD
   END (* swap *)
(* Now A <= B and B <= C, as desired.                      *)
```

Ex. 4: In general, how many exchanges must you write to order the values of n variables?

As you can see, this technique for ordering values would be very cumbersome for large numbers of values. That problem will be solved later. For now, review these algorithms to see how to find the larger of two values and how to order several values from largest to smallest.

BASIC INPUT AND OUTPUT

Most interesting programs use loops to manipulate large amounts of data. This section shows three kinds of loops for getting (and displaying) a quantity of data.

Assume that we have a constant and variables like these:

```
CONST
   SENTINEL = ...;          (*      (To be explained.)          *)

VAR
   DATUM : INTEGER;         (* One input value.                  *)
   N : INTEGER;             (* Number of values to input.        *)
   I : INTEGER;             (* Loop index.                       *)
```

Assume that the input is a sequence of integers, one integer per line. If the lines of input could be counted beforehand, we could put that number as an *extra* first input number, using it to get the rest:

```
(* Find out how much input to get:                               *)
READ( N );

(* Get and echo the N data values:                               *)
FOR I:=1 TO N DO
   BEGIN (* each value *)
   READLN( DATUM );
   WRITELN( DATUM )
   END (* each value *)
```

†*Ex. 5: What is a major disadvantage of this method?*

Rather than count the data, we could put an extra, but unusual, value as the *last* data item (called a *sentinel* value), then input everything up to that value. Assume that SENTINEL was set to that special value:

```
(* Get and echo data up to the sentinel (SENTINEL) value: *)
READLN( DATUM );
WHILE DATUM <> SENTINEL DO
   BEGIN (* each value *)
   WRITELN( DATUM );
   READLN( DATUM )
   END (* each value *)
```

*Ex. 6: What if all integer values are possible input values (that is, none is unusual enough to be used as a sentinel); can you use "END" or "***" as a sentinel?*

†*Ex. 7: You can use an unusual sequence of two values as a sentinel even if any single value is valid data. Write the algorithm for this, remembering that the first value of the pair is part of the sentinel if and only if the proper second value follows it—otherwise it is data. (Hint: include a "shift" algorithm.)*

The third way to get an input sequence is the easiest and most commonly used:

```
(* Get and echo data values until end-of-file:                    *)
WHILE NOT EOF(INPUT) DO
   BEGIN (* each value *)
   READLN( DATUM );
   WRITELN( DATUM )
   END (* each value *)
```

Ex. 8: In which of the three input loop control techniques can READ be substituted for READLN?

†*Ex. 9: Rewrite the EOF loop assuming that any number of values can appear on one line.*

There are many ways to vary and to combine these three input techniques. For example, the input sequence might be divided into smaller groups by one sentinel value (say, 0), with the whole sequence ended by another (maybe -1), like this:

```
 1  2  1  0
 4  0
 3  1  4  2  0
-1
```

Let GROUPEND be 0, and let SENTINEL be -1. (Actual data will be any positive integers.) The data might be processed this way:

```
(* Get and echo groups of data up to the final sentinel:   *)
READ( DATUM );
WHILE DATUM <> SENTINEL DO
   BEGIN (* each group *)

   (* Get and echo group values until the end of group:    *)
   WHILE DATUM <> GROUPEND DO
      BEGIN (* each value *)
      WRITE( DATUM );
      READ( DATUM )
      END; (* each value *)
   WRITELN;
   READ( DATUM )
   END (* each group *)
```

Ex. 10: Does this algorithm handle empty groups or a lack of groups properly?

†*Ex. 11: Rewrite the algorithm to use the EOF function instead of using the -1 sentinel value. (Take care about the ends of lines.)*

BASIC OPERATIONS ON DATA

In this section we explore some of the general kinds of operations that are commonly performed on sequences of input data.

Assume that we have these declarations:

```
VAR
   DATUM : INTEGER;        (* One input value.                    *)
   N : INTEGER;            (* Number of input values.             *)
   SUM : INTEGER;          (* Sum of the input values.            *)
   SMALL : INTEGER;        (* Smallest input value.               *)
```

Assume also that the data appear one value per line. EOF loops will be used throughout this section to get the data.

The list of values would usually be echoed in a single column. Sometimes, however, the single input sequence is divided between two output columns, using some criterion to decide the column in which each value will appear. Here negative versus nonnegative will be used as an example criterion to select a column:

```
(* Echo input data to selected output columns:                  *)

(* Print column headings:                                       *)
WRITELN( 'Negative':10, '0 or Positive':15 );

(* Get and echo data until end-of-file, putting the             *)
(* negative and nonnegative values into separate columns:       *)
WHILE NOT EOF(INPUT) DO
   BEGIN (* each value *)
   READLN( DATUM );
   (* Select appropriate output column, then echo:              *)
   IF DATUM < 0 THEN
      WRITELN( DATUM:10 )
   ELSE
      WRITELN( ' ':10, DATUM:15 )
   END (* each value *)
```

†*Ex. 12: Write an algorithm to select one of three columns, given an appropriate selection criterion.*

One reason for dividing the values among columns is to make the presence of certain values more obvious. This is often done by merely "marking" certain values. Using the preceding selection criterion again:

```
(* Get and echo data, marking nonnegative values:          *)
WHILE NOT EOF(INPUT) DO
   BEGIN (* each value *)
   READLN( DATUM );
   WRITE( DATUM );

   (* Select and mark nonnegatives:                        *)
   IF DATUM >= 0 THEN
      WRITE( '***' );

   (* Finish the line in any case:                         *)
   WRITELN
   END; (* each value *)

(* Explain the marking to the reader:                      *)
WRITELN;
WRITELN( '(Nonnegative values are marked with "***".)' )
```

Ex. 13: Rewrite the algorithm so that negatives are marked on the left, positives on the right, and zeros left unmarked. Keep the column of numbers straight.

Two of the most frequent operations on data are counts and summations. Both can be done at once:

```
(* Get and echo data until EOF, finding the sum (SUM) and *)
(* the number of items (N):                               *)
N := 0;
SUM := 0;
WHILE NOT EOF(INPUT) DO
   BEGIN (* each value *)

   (* Get and echo the item:                              *)
   READLN( DATUM );
   WRITELN( DATUM );

   (* Include this item in the count:                     *)
   N := N + 1;

   (* Include its value in the sum:                       *)
   SUM := SUM + DATUM;
   END; (* each value *)

(* Show the count and the sum:                            *)
WRITELN;
WRITELN( 'There are ', N:1, ' values whose sum is ', SUM:1 )
```

Ex. 14: How would you change the statements in the loop to print each value with its item number?

†*Ex. 15: In the previous section there is an algorithm (using two different sentinels) to input several groups of integers. Rewrite it to count the number of items in each group and the number of groups.*

Another important algorithm finds the smallest (or largest, or both) value in the input. Recall the simple algorithm that compares what is thought to be the smallest value with a new value, saving the new one if it is smaller:

```
(* Save the value of A if it is smaller than SMALL:          *)
IF A < SMALL THEN
   SMALL := A
```

To find the smallest overall value, this simple algorithm is merely performed upon each and every input value (starting with an impossibly large value for SMALL):

```
(* Get and echo data until EOF, setting SMALL to the          *)
(* smallest value encountered in the data:                    *)
SMALL := MAXINT;
WHILE NOT EOF(INPUT) DO
   BEGIN (* each value *)
   READLN( DATUM );
   WRITELN( DATUM );

   (* Save DATUM if it is the smallest so far:                *)
   IF DATUM < SMALL THEN
      SMALL := DATUM
   END; (* each value *)

WRITELN( 'The smallest value in the data is ', SMALL:1 )
```

Ex. 16: Why was SMALL initialized to MAXINT rather than, for instance, to 0? What is the result if there is no input?

†*Ex. 17: Assume that MAXINT is not available, that you do not know its value, and that the input can include any integers. Rewrite the algorithm to find the smallest input value. (Hint: SMALL can only be set to actual input values.)*

This is called a *search* algorithm; a sequence of values is searched (inspected one at a time) to find a value with a particular property or magnitude.

The algorithm that marked nonnegative values was in fact a search that found many values with the desired property. Often we are more interested in the several values found by such a search than in the data as a whole. For example, we may want to count and sum only the positive input values:

```
(* Get and echo data, counting and summing the positive    *)
(* values only:                                             *)
N := 0;
SUM := 0;
WHILE NOT EOF(INPUT) DO
   BEGIN (* each value *)
   READLN( DATUM );
   WRITELN( DATUM );

   (* Find and process positive values:                    *)
   IF DATUM > 0 THEN
      BEGIN (* another positive *)
      N := N + 1;
      SUM := SUM + DATUM
      END (* another positive *)
   END (* each value *)

(* Report the outcome:                                      *)
WRITELN;
WRITELN( 'The ', N:1, ' positive values totaled ', SUM:1 )
```

Ex. 18: Rewrite this algorithm to count and sum the positive and the negative values separately and to count the zero values.

PROGRAMS FOR READING

```
 1  PROGRAM  PAY( INPUT, OUTPUT );
 2
 3  (* PURPOSE:                                                   *)
 4  (*    To compute pay raises for a group of employees and     *)
 5  (*    print a report.                                         *)
 6  (*                                                            *)
 7  (* PROGRAMMER:  John Potok                                    *)
 8  (*                                                            *)
 9  (* INPUT:                                                     *)
10  (*    One line of data for each employee.  Each line          *)
11  (*    contains an ID number, a space, and then the            *)
12  (*    current wage in the form ddd.cc (dollars and cents).*)
13  (*    (The rate of wage increase is a program constant.)     *)
14  (*                                                            *)
15  (* OUTPUT:                                                    *)
16  (*    A table of old and new wages, followed by a summary,*)
17  (*    including maximum and minimum wages (after the          *)
18  (*    raises).                                                *)
19  (*                                                            *)
20  (* ASSUMPTIONS/LIMITATIONS:                                   *)
21  (*    The ID numbers may have up to 9 digits.                 *)
22  (*    All data must be correct and in correct order.          *)
23  (*                                                            *)
24  (* ERROR CHECKS & RESPONSES:                                  *)
25  (*    None.                                                   *)
26  (*                                                            *)
27  (* ALGORITHM/STRATEGY:                                        *)
28  (*    Initialize counter, minimum and maximum.                *)
29  (*    Print column headings.                                  *)
30  (*    Process employee wages until EOF:                       *)
31  (*       Get and echo ID and current wage.                    *)
32  (*       Calculate and print the new wage.                    *)
33  (*       Save the minimum and maximum so far.                 *)
34  (*    Print the summary.                                      *)
35
36  CONST
37     PAYRAISE = 0.20;          (* A 20 percent rate increase. *)
38
39  VAR
40     IDNUM : INTEGER;          (* The employee ID number.     *)
41     EMPLOYEES : INTEGER;      (* The number of employees.    *)
42     OLDWAGE : REAL;           (* Employee's old wage rate.   *)
43     NEWWAGE : REAL;           (* Employee's new wage rate.   *)
44     MINWAGE : REAL;           (* The minimum wage rate.      *)
45     MAXWAGE : REAL;           (* The maximum wage rate.      *)
46
```

```
47   BEGIN (* PAY *)
48
49   (* Initialize the employee counter, minimum and maximum:   *)
50   EMPLOYEES := 0;
51   MINWAGE := 9999.99;
52   MAXWAGE := -9999.99;
53
54   (* Print column headings for the report:                   *)
55   PAGE(OUTPUT);
56   WRITELN( '**** TABLE OF WAGE INCREASES ****':41 );
57   WRITELN;
58   WRITELN;
59   WRITELN( 'ID NUMBER':13, 'OLD WAGE RATE':17,
60            'NEW WAGE RATE':17 );
61
62   (* Process employee wages until EOF:                       *)
63   WHILE NOT EOF(INPUT) DO
64      BEGIN (* each employee *)
65
66      (* Get and echo the ID number and current wage:         *)
67      READLN( IDNUM, OLDWAGE );
68      WRITE( IDNUM:12, OLDWAGE:12:2, ' / hr' );
69
70      (* Calculate the new wage and count this employee:      *)
71      NEWWAGE := OLDWAGE + OLDWAGE*PAYRAISE;
72      EMPLOYEES := EMPLOYEES + 1;
73
74      (* Report the new wage:                                 *)
75      WRITELN( NEWWAGE:12:2, ' / hr' );
76
77      (* Save the value of the smallest wage rate in MINWAGE *)
78      (* and largest rate in MAXWAGE:                         *)
79      IF NEWWAGE < MINWAGE THEN
80         MINWAGE := NEWWAGE
81      ELSE IF NEWWAGE > MAXWAGE THEN
82         MAXWAGE := NEWWAGE
83
84      END; (* each employee *)
85
86   (* Report the number of persons receiving wage increases, *)
87   (* the maximum wage rate and the minimum wage rate:        *)
88   WRITELN;
89   WRITELN( 'NUMBER OF EMPLOYEES:':30, EMPLOYEES:3 );
90   WRITELN( 'LARGEST WAGE RATE:':30, MAXWAGE:6:2, ' *' );
91   WRITELN( 'SMALLEST WAGE RATE:':30, MINWAGE:6:2, ' *' );
92   WRITELN;
93   WRITELN;
94   WRITELN( '*  AFTER THE WAGE INCREASE.':38 );
95   WRITELN;
96   WRITELN;
97   WRITELN( '*** END OF REPORT ***':33 );
98   PAGE(OUTPUT)
99   END. (* PAY *)
```

```
     **** TABLE OF WAGE INCREASES ****

ID NUMBER     OLD WAGE RATE     NEW WAGE RATE
 1981926         3.50 / hr         4.20 / hr
 6151196         9.95 / hr        11.94 / hr
 4249400         5.45 / hr         6.54 / hr
 4217509         6.25 / hr         7.50 / hr
 9811150         4.05 / hr         4.86 / hr
 2765085         7.35 / hr         8.82 / hr

      NUMBER OF EMPLOYEES:  6
        LARGEST WAGE RATE: 11.94 *
       SMALLEST WAGE RATE:  4.20 *


       *  AFTER THE WAGE INCREASE.


        *** END OF REPORT ***
```

```
 1  PROGRAM CHECKACCT( INPUT, OUTPUT );
 2
 3  (* PURPOSE:                                                *)
 4  (*     To generate a simple checking account statement.    *)
 5  (*                                                         *)
 6  (* PROGRAMMER: John Potok                                  *)
 7  (*                                                         *)
 8  (* INPUT:                                                  *)
 9  (*     A sequence of debits (checks and service charges)   *)
10  (*     and credits (deposits) to the account, preceded by  *)
11  (*     a balance.  Each transaction has this format:       *)
12  (*          column 1:  B for balance (first line only)     *)
13  (*                     C for check                         *)
14  (*                     S for service charge                *)
15  (*                     D for deposit                       *)
16  (*          column 2:  blank                               *)
17  (*      columns 3-10:  amount ($) in the form ddddd.cc     *)
18  (*                                                         *)
19  (* OUTPUT:                                                 *)
20  (*     An itemized statement of transactions on the        *)
21  (*     account, and an overall summary.                    *)
22  (*                                                         *)
23  (* ASSUMPTIONS/LIMITATIONS:                                *)
24  (*     The balance (B) transaction should be first.        *)
25  (*                                                         *)
26  (* ERROR CHECKS & RESPONSES:                               *)
27  (*     A zero or negative balance is flagged with "****".  *)
28  (*                                                         *)
29  (* ALGORITHM/STRATEGY:                                     *)
30  (*     Initialize counts and sums:                         *)
31  (*     Print column headings                               *)
32  (*     Read and process the transactions until EOF:        *)
33  (*        Get a transaction                                *)
34  (*        Perform an action depending on transaction type: *)
35  (*           B--set and report balance                     *)
36  (*           C--handle debit for a check                   *)
37  (*           S--handle debit for a service charge          *)
38  (*           D--credit a deposit to the account            *)
39  (*        Mark an overdraft, if necessary.                 *)
40  (*     Report debit and credit totals                      *)
41  (*     Report number of transactions                       *)
42
43  VAR
44     TRANSACTION : CHAR;        (* The type of transaction.    *)
45     AMOUNT : REAL;             (* The dollar amount of        *)
46                                (* a particular transaction.   *)
47     NUMDEPOSITS : INTEGER;     (* Number of deposits.         *)
48     NUMCHECKS : INTEGER;       (* Number of checks.           *)
49     NUMSERVICES : INTEGER;     (* Number of service charges.  *)
50     CREDITS : REAL;            (* The total amount of credits.*)
51     DEBITS : REAL;             (* The total amount of debits. *)
52     BALANCE : REAL;            (* The running balance.        *)
53
```

```
 54  BEGIN (* CHECKACCT *)
 55
 56  (* Initialize counts and sums:                              *)
 57  NUMDEPOSITS := 0;
 58  NUMCHECKS := 0;
 59  NUMSERVICES := 0;
 60  CREDITS := 0.0;
 61  DEBITS := 0.0;
 62
 63  (* Print column headings for checking account statement:  *)
 64  PAGE(OUTPUT);
 65  WRITELN( '---STATEMENT OF ACCOUNT--':34 );
 66  WRITELN;
 67  WRITELN( 'ITEM         ', 'CREDITS':10, 'DEBITS':10,
 68           'BALANCE':10 );
 69  WRITELN;
 70
 71  (* Read and process the transactions until EOF:             *)
 72  WHILE NOT EOF(INPUT) DO
 73     BEGIN (* each TRANSACTION *)
 74
 75     (* Get the transaction:                                   *)
 76     READLN( TRANSACTION, AMOUNT );
 77
 78     (* Depending upon the type of transaction, adjust        *)
 79     (* BALANCE and report the transaction:                    *)
 80     CASE TRANSACTION OF
 81
 82        'B': BEGIN (* balance *)
 83             (* BALANCE is assigned the beginning balance:  *)
 84             BALANCE := AMOUNT;
 85             (* Report the beginning balance:                 *)
 86             WRITE( 'balance      ', AMOUNT:10:2, ' ':10,
 87                    BALANCE:10:2 )
 88             END; (* balance *)
 89
 90        'C': BEGIN (* check *)
 91             (* Find the total debits, the new balance, and *)
 92             (* the number of checks so far:                  *)
 93             DEBITS := DEBITS + AMOUNT;
 94             BALANCE := BALANCE - AMOUNT;
 95             NUMCHECKS := NUMCHECKS + 1;
 96             (* Report the check and the new balance:         *)
 97             WRITE( 'check        ', ' ':10, AMOUNT:10:2,
 98                    BALANCE:10:2 )
 99             END; (* check *)
100
```

```
101        'D': BEGIN (* deposit *)
102             (* Find the total credits, the new balance,     *)
103             (* and the number of deposits so far:           *)
104             CREDITS := CREDITS + AMOUNT;
105             BALANCE := BALANCE + AMOUNT;
106             NUMDEPOSITS := NUMDEPOSITS + 1;
107             (* Report the deposit and the new balance:      *)
108             WRITE( 'deposit     ', AMOUNT:10:2, ' ':10,
109                    BALANCE:10:2 )
110             END; (* deposit *)
111
112        'S': BEGIN (* service charge *)
113             (* Find the total service charges, the new      *)
114             (* balance and the number of service charges    *)
115             (* so far:                                      *)
116             DEBITS := DEBITS + AMOUNT;
117             BALANCE := BALANCE - AMOUNT;
118             NUMSERVICES := NUMSERVICES + 1;
119             (* Report the service charges:                  *)
120             WRITE( 'serv. charge', ' ':10, AMOUNT:10:2,
121                    BALANCE:10:2 )
122             END; (* service charges *)
123
124     END; (* TRANSACTION cases *)
125
126     (* Select and mark any overdrafts:                      *)
127     IF BALANCE <= 0.0 THEN
128        WRITE( '****':6 );
129
130     WRITELN
131     END; (* each TRANSACTION *)
132
133  (* Report the debit and credit totals:                     *)
134  WRITELN( ' ':12, '-------':10, '-------':10 );
135  WRITELN( 'TOTALS      ', CREDITS:10:2, DEBITS:10:2,
136           BALANCE:10:2 );
137
138  (* Report the total number of each kind of transaction:    *)
139  WRITELN;
140  WRITELN;
141  WRITELN( 'Deposits:':16, NUMDEPOSITS:4 );
142  WRITELN( 'Checks:':16, NUMCHECKS:4 );
143  WRITELN( 'Service Charges:':16, NUMSERVICES:4 );
144  WRITELN;
145  WRITELN;
146
147  (* Explain the marking to the reader:                      *)
148  WRITELN( 'Overdrafts are marked with "****".' );
149  WRITELN;
150  WRITELN;
151  WRITELN( '---END OF STATEMENT--':32 );
152  PAGE(OUTPUT)
153  END. (* CHECKACCT *)
```

```
        ---STATEMENT OF ACCOUNT--

ITEM              CREDITS     DEBITS    BALANCE

balance            291.86                291.86
check                          10.66     281.20
check                         200.00      81.20
check                          92.81     -11.61   ****
check                          15.30     -26.91   ****
deposit            241.86                214.95
serv. charge                    3.74     211.21
check                         122.00      89.21
check                           7.98      81.23
check                          29.21      52.02
check                          16.64      35.38
check                          39.40      -4.02   ****
deposit            246.12                242.10
check                          39.40     202.70
                  -------    -------
TOTALS             487.98     577.14     202.70


       Deposits:    2
         Checks:   10
Service Charges:    1


Overdrafts are marked with "****".


          ---END OF STATEMENT--
```

PROGRAM EXERCISES

Note: Whenever you are asked to show changes to a program, you should indicate changed, inserted, or deleted lines this way:

> change: 38 ...changed line...
> insert: 65.1 ...new line...
> insert: 65.2 ...new line...
> delete: 80

1. (a) Find all the statements in program PAY that are part of the algorithms that find the smallest and largest wages in the data.
   (b) Show the output corresponding to this input:

```
1111111    5.00
```

   (c) Show the output that corresponds to this input:

```
1111111    6.00
2222222    5.00
3333333    4.00
```

   (d) Show the changes to PAY that will make it produce correct results for the two sets of data shown above.

2. Show all the necessary changes to program PAY so that it will find and print the increase in total hourly wages.

3. Show the changes to CHECKACCT that will make it accomplish one or both of these "self-defense" strategies:

   (a) Refuse to process a balance ("B") transaction that is not the first input line.
   (b) Refuse to process (but print an error message for) any TRANSACTION that is not a "B", "C", "D", or "S".

4. Are any of the following lines of data unacceptable to the PAY program?

```
1234567      1.98
    99     100.22
      1       1000.00
```

5. (a) Are any of the following transactions for CHECKACCT invalid?

```
D           207.05
 C3.98
      C      12.87
```

   (b) Show how to change the program so that all these transactions are acceptable.

Part II: Character-Processing Algorithms

INTRODUCTION

Character processing represents a distinct class of widely used algorithms. Both individually and in combination with string processing (discussed in Part V), these algorithms are used in many programs, but especially in compilers, text formatters, operating systems, and communications.

Characters can be processed as individual, essentially unrelated values. This simplistic approach is obviously inadequate, however, because character data is almost always a representation of natural or formal languages. It is conceptually divided into words and sentences, identifiers and statements. Certain character values, by tradition or by design, have special consistent meanings. Blanks separate words, for example. The fact that character data represents language means, in essence, that characters often are of greater interest in combination than individually. This fact has implications for input and output and for whatever happens in between.

BASIC INPUT AND OUTPUT

It is important to see first how character data may be treated like any other sequence of values. Some elements of this treatment will always be a part of character-processing programs.

Assume that this variable has been declared:

```
VAR
   CH : CHAR;              (* One character of data.          *)
```

One technique for getting textual data is quite similar to the methods for numeric data:

```
(* Get and echo all lines of text:                            *)
WHILE NOT EOF(INPUT) DO
   BEGIN (* each line *)

   (* Get and echo all characters in the line:                *)
   WHILE NOT EOLN(INPUT) DO
      BEGIN (* each char *)
      READ( CH );
      WRITE( CH )
      END; (* each char *)

   (* Finish the line:                                        *)
   READLN;
   WRITELN

   END (* each line *)
```

Character data is also often counted. Assume that these variables are also available:

```
CHPERLINE : INTEGER; (* Characters on one input line.  *)
TOTALCH : INTEGER;   (* Total number of characters.    *)
LINES : INTEGER;     (* Number of lines of input.      *)
```

Note that although this algorithm deals with characters, its organization applies to any sequence of data that is organized into shorter groups separated by sentinel values:

```
(* Get and echo all lines of text, counting characters      *)
(* per line, total characters, and number of lines:         *)
TOTALCH := 0;
LINES := 0;
WHILE NOT EOF(INPUT) DO
   BEGIN (* each line *)

   (* Get and echo the chars on a line, counting them:      *)
   CHPERLINE := 0;
   WHILE NOT EOLN(INPUT) DO
      BEGIN (* each char *)
      READ( CH );
      WRITE( CH );
      (* Count this character:                              *)
      CHPERLINE := CHPERLINE + 1
      END; (* each char *)

   (* Show the count, then finish the line:                 *)
   WRITELN( '    (', CHPERLINE:1, ' characters)' );
   READLN;

   (* Include this line's characters in the total:          *)
   TOTALCH := TOTALCH + CHPERLINE;

   (* Count this line:                                      *)
   LINES := LINES + 1

   END;   (* each line *)

(* Show the total and line count:                           *)
WRITELN;
WRITELN( 'There are ', LINES:1, ' lines containing ',
         TOTALCH:1, ' characters overall.' )
```

A little reflection reveals that EOF and EOLN in effect test for *sentinel values* that are provided by the language. These special values will never appear as actual data.

TESTS AND TRANSLATIONS

When processing characters, it is usually necessary to find the "class" to which a character belongs. The common classes are the alphabetic characters, the digits, uppercase alphabetics, and so on. Classes are just groups of related characters. It is also often necessary to translate characters (change their values), usually from one class to another.

Assume that we have these declarations:

```
VAR
   UPTOLOW : INTEGER;          (* "Distance" between the ordinal *)
                               (* values of upper- and lowercase *)
                               (* characters.                    *)
   CH : CHAR;                  (* One input character.           *)
```

where UPTOLOW will be assigned the numeric quantity ORD('A') - ORD('a'). This quantity varies from one character set to another.

The EBCDIC character set has the unfortunate property that the letters of the alphabet are not arranged in a sequence of 26 contiguous characters. There are, in fact, extra characters between 'I' and 'J' and between 'R' and 'S'. (Similarly for lowercase letters.) Fortunately, these extra characters will seldom be encountered in the input stream. (There are no such extra characters in the ASCII character set used on most computers.) For the sake of simplicity, the letters are considered contiguous here.

A character can be tested to see if it lies within a range of values:

```
(* See if CH is an alphabetic or a digit:                    *)
IF ('A' <= CH) AND (CH <= 'Z') THEN
   WRITELN( CH, ' is a capital letter.' )
ELSE IF ('a' <= CH) AND (CH <= 'z') THEN
   WRITELN( CH, ' is a lowercase letter.' )
ELSE IF ('0' <= CH) AND (CH <= '9') THEN
   WRITELN( CH, ' is a digit.' )
```

The same tests can be performed much more succinctly using set constants:

```
(* Test CH and report its class:                             *)
WRITE( CH, ' is ' );
IF CH IN ['A'..'Z'] THEN
   WRITELN( 'uppercase.' )
ELSE IF CH IN ['a'..'z'] THEN
   WRITELN( 'lowercase.' )
ELSE IF CH IN ['0'..'9'] THEN
   WRITELN( 'a digit.' )
```

(Sets have many interesting properties, but they are used in this book only to check the values of variables. The maximum size of a set is implementation dependent, but the standard for Pascal shows that they should be at least large enough to include all characters in the character set.)

Set constants would also be used to test a character against several discrete values or ranges:

```
(* See which classes CH belongs to:                                *)
IF CH IN ['A'..'Z', 'a'..'z'] THEN
   WRITELN( CH, ' is an alphabetic.' );
IF CH IN ['A','E','I','O','U','a','e','i','o','u'] THEN
   WRITELN( CH, ' is a vowel.' );
IF CH IN ['.', ',', ':', ';', '!', '?'] THEN
   WRITELN( '"', CH, '" is punctuation.' )
```

Finally, we will often want to *translate* one character into another:

```
(* Translate CH from uppercase to lowercase:                       *)
CH := CHR( ORD(CH) - UPTOLOW )
```

or any of several characters into one:

```
(* "Delete" any punctuation before using CH:                       *)
(* (Change any punctuation character to a blank.)                  *)
IF CH IN ['.', ',', ':', ';', '!', '?'] THEN
   CH := ' '
```

or a character digit into its numeric value:

```
(* Translate the digit CH into an integer I:                       *)
I := ORD(CH) - ORD('0')
```

The reasons for doing these translations will become apparent in later sections.

Ex. 1: Write tests for end-of-sentence punctuation and for middle-of-word punctuation.

Ex. 2: Given a pair of character digits named TENS and ONES, convert the pair into a two-digit integer.

†*Ex. 3: Write an algorithm to input a sequence of character digits, converting them into one integer value.*

STRATEGIES: SENTINELS VERSUS TRANSITIONS

There are two general strategies for devising character-processing algorithms—one using character sentinel values to control loops, the other using character class "transitions" to make processing decisions. These methods are so general that we will have to use concrete examples to explain them.

Assume that we have this variable:

```
VAR
   CH : CHAR;                (* One character of input.          *)
```

Let us say that we want to input a stream of text and print one word of it per line of output. Assume that there is one blank (or EOLN, which reads as a blank) after each word. Using what we already know about loop control, we can view the text as groups of nonblanks (words), each terminated by a blank sentinel value:

```
(* Get text until EOF, printing one word per line:           *)
WHILE NOT EOF(INPUT) DO
   BEGIN (* each word *)

   (* Get and echo the next word, up to a blank:             *)
   READ( CH );
   WHILE CH <> ' ' DO
      BEGIN (* each nonblank *)
      WRITE( CH );
      READ( CH )
      END; (* each nonblank *)

   (* Finish the line:                                       *)
   WRITELN

   (*  (We have already read the trailing blank.)            *)
   END (* each word *)
```

To use the transition strategy, note that blanks occur *only* after words, so that any time a blank is encountered (and only then) it signals the end of a word of input and a line of output:

```
(* Get text until EOF, printing one word per line:           *)
WHILE NOT EOF(INPUT) DO
   BEGIN (* each character *)

   (* Read it:                                                *)
   READ( CH );

   (* Decide what action to take:                             *)
   IF CH = ' ' THEN
      WRITELN
   ELSE (* CH <> ' ' *)
      WRITE( CH )
   END (* each character *)
```

The actions taken by this algorithm, given a character, can be summarized in a very simple table:

```
CHARACTER:  blank       nonblank
   ACTION:  new line    echo it
```

If the problem is extended slightly so that *one or more* blanks are allowed after each word, we can use blanks and nonblanks as sentinels for each other. The first example now becomes:

```
(* Get text until EOF, printing one word per line:           *)
READ( CH );
WHILE NOT EOF(INPUT) DO
   BEGIN (* each word *)

   (* Get and echo the rest of the word, up to a blank:      *)
   WHILE CH <> ' ' DO
      BEGIN (* each nonblank *)
      WRITE( CH );
      READ( CH )
      END; (* each nonblank *)
   WRITELN;

   (* Skip the blanks, if any, up to a nonblank or EOF:      *)
   WHILE (CH = ' ') AND NOT EOF(INPUT) DO
      READ( CH )

   (* (We now might have the 1st letter of the next word) *)
   END (* each word *)
```

†*Ex. 4: Why is there a READ before the loop? Why is there an extra EOF test inside the loop?*

Ex. 5: Rewrite the algorithm to allow blanks before the first word.

To rewrite this new algorithm using the transition strategy, the input must be viewed differently. Looking at any single character of input, it is either in a word (it is nonblank) or it is in a sequence of blanks (it is blank). If it is in a word, it should be printed; otherwise it should be ignored. There are thus two processing situations or *states.* If the state is "in" a word and a blank is encountered, then the state must change to the blanks processing state, and vice versa. This change is called a *state transition* (or simply a transition). Since the program looks at only one character at a time, it must keep track of the state it is in while it gets the next character. We can simply use a boolean variable to represent the two states of this particular problem:

```
INAWORD : BOOLEAN;     (* TRUE if state is "in a word",  *)
                       (* FALSE if state is "in blanks". *)
```

Earlier we noted that the actions to be taken depended upon each character of input. This is still true, but the action will now also depend upon the state the program is in. For example, if it is in a word and it gets a blank, it should start a new line (and get into the blanks state), but if it is in the blanks state, it should ignore the blanks it gets. All the decisions to be made for this specific problem can be summarized in a *state-transition table* like this:

```
                              Input Character:
 State:                     a blank        a nonblank

in a word  (action:)*      new line       echo it
        (next state:)      in blanks      in a word

 in blanks  (action:)      ignore it      echo it
        (next state:)      in blanks      in a word

  *Assume that processing starts in this state.
```

We can show the same thing using a *state-transition diagram.* The arrows show the transitions from one state to another (or to the same state) given an input character. These arrows are labeled with "input/action" to show the processing of each character:

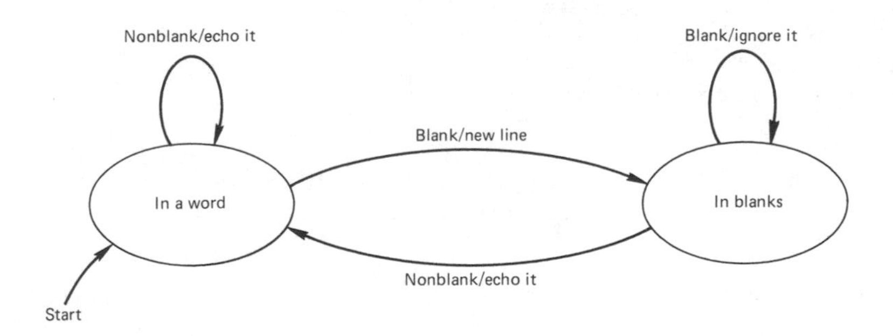

Ex. 6: Draw the state transition diagram for the original problem, where there was one blank after each word. (It uses only one state.)

Finally, the algorithm can be implemented this way, using the order of states and transitions shown in the table:

```
(* Get text until EOF, printing one word per line:        *)
INAWORD := TRUE;
WHILE NOT EOF(INPUT) DO
   BEGIN (* each char *)

   (* Get the character:                                   *)
   READ( CH );

   (* Given the current state and the character,           *)
   (* determine the action and the new state:              *)
   IF INAWORD THEN
      BEGIN (* "in a word" state *)
      IF CH = ' ' THEN
         BEGIN (* found a blank *)
         WRITELN;
         (* Switch state to "in blanks":                   *)
         INAWORD := FALSE
         END (* found a blank *)
      ELSE (* CH <> ' ' *)
         WRITE( CH )
      END (* "in a word" state *)

   ELSE
      BEGIN (* "in blanks" state *)
      IF CH = ' ' THEN
         (* Ignore it--do nothing *)
      ELSE
         BEGIN (* found a nonblank *)
         WRITE( CH );
         INAWORD := TRUE
         END (* found a nonblank *)
      END (* state "in blanks" *)

   END (* each char *)
```

†*Ex. 7: What simple change prevents this algorithm from skipping an extra line if there are blanks before the first word?*

A review of the example algorithms in this section reveals that the state-transition versions are longer than the versions that use the sentinel strategy. This is often (but not always) the case. There are ways, however, to simplify and condense the state-transition algorithms, as you will see in the next section. In any case, this strategy has some very important properties: it is a straightforward, methodical technique; it gives the programmer greater control and thus is less subject to error; it results in algorithms that are more easily modified or augmented. These properties (and the size factor) are due mostly to the fact that each problem is expressed only in terms of inputs, actions, and state transitions—even when some parts of the problem might be done more briefly some other way. It also turns out that more complex text-processing problems can be done more easily and briefly this way than with the sentinel strategy.

CHARACTER CLASSES

As shown in the examples of the previous section, many character-processing algorithms are concerned more with classes of values than with specific values. That is, all alphabetics may be processed one way, all end-of-sentence punctuation another way, and so on. Such algorithms could be clearer if the character classes were made more obvious.

Assume that we represent some classes and states this way:

```
TYPE
   CHARCLASSES =           (* Character classes:              *)
      (BLANK,              (* Stands for the blank only.      *)
       NONBLANK);          (* Stands for all other characters.*)
   ALLSTATES =             (* Processing states:              *)
      (INAWORD,            (* Doing nonblanks.                *)
       INBLANKS);          (* Doing blanks.                   *)

VAR
   CH : CHAR;              (* One character of input.         *)
   CLASS : CHARCLASSES;(* CH's character class.               *)
   STATE : ALLSTATES;      (* The current processing state.   *)
```

When we get a character of input, we can also find its class:

```
(* Read the next character and determine its class:          *)
READ( CH );
IF CH = ' ' THEN
   CLASS := BLANK
ELSE
   CLASS := NONBLANK
```

In general, a separate class is necessary for each distinction to be made among characters (BLANK and NONBLANK) in the algorithms, just as a separate state is needed for each distinction made among processing situations (INAWORD and INBLANKS) in those algorithms.

†Ex. 8: Declare CHARCLASSES to distinguish among the various classes of characters used in programming.

The distinctions made among characters (and states) depend in turn upon how they are to be processed.

Now the simple text-processing problem of the previous section can be restructured to use classes, the clearer enumerated types, and the CASE statement (a simpler control construct):

```
(* Get text until EOF, printing one word per line:        *)
STATE := INAWORD;
WHILE NOT EOF(INPUT) DO
   BEGIN (* each char *)

   (* Get the next character and find its class:          *)
   READ( CH );
   IF CH = ' ' THEN
      CLASS := BLANK
   ELSE
      CLASS := NONBLANK;

   (* Given the current state and the class, determine    *)
   (* the action and the new state:                        *)
   CASE STATE OF

      INAWORD:
         CASE CLASS OF
            NONBLANK:
               WRITE( CH );
            BLANK:
               BEGIN (* 1st blank *)
               WRITELN;
               STATE := INBLANKS
               END (* 1st blank *)
         END; (* CLASS cases *)

      INBLANKS:
         CASE CLASS OF
            NONBLANK:
               BEGIN (* 1st nonblank *)
               WRITE( CH );
               STATE := INAWORD
               END; (* 1st nonblank *)
            BLANK:
               (* Ignore it--do nothing *)
         END (* CLASS cases *)

   END (* STATE cases *)

   END (* each char *)
```

Typically the character is read, and its class assigned, in a procedure. In its simplest form the procedure would look like this:

```
PROCEDURE  GETNEXT( VAR CH: CHAR; VAR CLASS: CHARCLASSES );
   (* Read CH, the next character of input, and find its  *)
   (* CLASS, returning both values to the caller:          *)

   BEGIN (* GETNEXT *)

   READ( CH );
   IF CH = ' ' THEN
      CLASS := BLANK
   ELSE
      CLASS := NONBLANK

   END; (* GETNEXT *)
```

A more typical GETNEXT procedure would perform more functions. It could echo the input to a special file, or translate letters from uppercase to lowercase, for example. The character classes could be expanded to include, say, ENDFILE and ENDLINE, and the procedure would test for these, doing what it needed to do for the echo, and passing the information back to the state-transition algorithm.

Ex. 9: Change the class declaration, the GETNEXT procedure, and the transitions in the simple text-processing algorithm so that '*****' *is printed on a new line when any end-of-sentence punctuation is encountered.*

PROGRAMS FOR READING

```
 1   PROGRAM  TRAILTEXT( INPUT, OUTPUT );
 2
 3   (* PURPOSE:                                                  *)
 4   (*    Compute statistics about the characters, words, and *)
 5   (*    sentences in some text, using the sentinel strategy.*)
 6   (*                                                           *)
 7   (* PROGRAMMER: John Potok                                    *)
 8   (*                                                           *)
 9   (* INPUT:                                                    *)
10   (*    One or more lines of text to be analyzed.              *)
11   (*                                                           *)
12   (* OUTPUT:                                                   *)
13   (*    An echo of the input text.                             *)
14   (*    A summary showing the number of characters,            *)
15   (*    words, and sentences that the text contains,           *)
16   (*    and the largest and average word and sentence.         *)
17   (*                                                           *)
18   (* ASSUMPTIONS/LIMITATIONS:                                  *)
19   (*    The input text is correctly punctuated.                *)
20   (*    Periods, question marks and exclamation points only *)
21   (*    occur in the last word of a sentence.                  *)
22   (*    Blanks separate words.                                 *)
23   (*                                                           *)
24   (* ERROR CHECKS & RESPONSES:                                 *)
25   (*    None.                                                  *)
26   (*                                                           *)
27   (* ALGORITHM/STRATEGY:                                       *)
28   (*    Initialize counters and set the first character        *)
29   (*       to a blank.                                         *)
30   (*    Print a heading for the echo of the text.              *)
31   (*    Process words until EOF.                               *)
32   (*       Step through sentinels between words.               *)
33   (*       Count first character of word.                      *)
34   (*       Step through rest of word until sentinel found:     *)
35   (*          Count the letters.                               *)
36   (*       Update word statistics.                             *)
37   (*       Check for end-of-sentence, update sentence stats.*)
38   (*       Process EOLN, if necessary.                         *)
39   (*    Compute averages of characters per word and            *)
40   (*       words per sentence.                                 *)
41   (*    Report statistical information.                        *)
42
```

```
43  VAR
44     CH : CHAR;                  (* One input character.          *)
45     CHARSINWORD : INTEGER;      (* # of characters in a word.    *)
46     WORDSINSENT : INTEGER;      (* # of words in a sentence.     *)
47     TOTALCHARS: INTEGER;        (* The number of nonsentinel     *)
48                                 (* characters in the text.       *)
49     TOTALWORDS : INTEGER;       (* Total # of words in text.     *)
50     TOTALSENTS : INTEGER;       (* Total # of sentences in text*)
51     LONGWORD : INTEGER;         (* # of chars in longest word. *)
52     LONGSENT : INTEGER;         (* # of words in longest sent. *)
53     AVGCHAR : REAL;             (* Ave. # of chars in a word.    *)
54     AVGWORD : REAL;             (* Ave. # of words in a sent.    *)
55
56
57
58
59
60  BEGIN (* TRAILTEXT *)
61
62  (* Initialize variables:                                          *)
63  CHARSINWORD := 0;
64  WORDSINSENT := 0;
65  TOTALCHARS:= 0;
66  TOTALWORDS := 0;
67  TOTALSENTS := 0;
68  LONGWORD := 0;
69  LONGSENT := 0;
70
71  (* Print a heading for the echo:                                  *)
72  PAGE( OUTPUT );
73  WRITELN( 'Input text:':29 );
74  WRITELN;
75
76  (* Set CH as if the first line begins with a blank:               *)
77  CH := ' ';
78
```

```
 79  (* Process words until EOF:                                   *)
 80  WHILE NOT EOF(INPUT) DO
 81     BEGIN (* Each word *)
 82
 83     (* Step through sentinels between words:                   *)
 84     WHILE CH IN [ '.', '!', '?', ' ' ] DO
 85        BEGIN (* Each sentinel *)
 86        READ( CH );
 87        WRITE( CH )
 88        END; (* Each sentinel *)
 89
 90     (* This char is either a letter or normal punctuation: *)
 91     IF NOT( CH IN [ ',', ':', ';' ] ) THEN
 92         CHARSINWORD := CHARSINWORD + 1;
 93
 94     (* Step through the characters of the word until any   *)
 95     (* end of sentence sentinel is found:                   *)
 96     WHILE NOT EOLN(INPUT)
 97           AND NOT (CH IN [ '.', '!', '?', ' ' ]) DO
 98        BEGIN (* Each non-sentinel *)
 99        (* Get and echo the next one:                           *)
100        READ( CH );
101        WRITE( CH );
102        (* Count the letters only:                              *)
103        IF CH IN [ 'A'..'Z', 'a'..'z' ] THEN
104            CHARSINWORD := CHARSINWORD + 1
105        END; (* Each non-sentinel *)
106
107     (* This is the end of a word.  Update counts and sum:   *)
108     WORDSINSENT := WORDSINSENT + 1;
109     TOTALCHARS := TOTALCHARS + CHARSINWORD;
110     (* See if this is the longest word:                        *)
111     IF LONGWORD < CHARSINWORD THEN
112        LONGWORD := CHARSINWORD;
113     (* Reinitialize for the next word:                         *)
114     CHARSINWORD := 0;
115
116     (* See if the last character was the end of sentence:   *)
117     IF CH IN [ '.', '!', '?' ] THEN
118        BEGIN (* Each sentence *)
119        TOTALSENTS := TOTALSENTS + 1;
120        TOTALWORDS := TOTALWORDS + WORDSINSENT;
121        (* See if this was the longest sentence:               *)
122        IF LONGSENT < WORDSINSENT THEN
123           LONGSENT := WORDSINSENT;
124        (* Reinitialize for the next sentence:                 *)
125        WORDSINSENT := 0
126        END; (* Each sentence *)
127
128     IF EOLN(INPUT) THEN
129        BEGIN (* at EOLN *)
130        READ( CH );
131        WRITELN
132        END (* at EOLN *)
133     END; (* Each word *)
```

```
134
135  (* Compute the average number of characters per word and   *)
136  (* average number of words per sentence:                   *)
137  AVGCHAR := TOTALCHARS / TOTALWORDS;
138  AVGWORD := TOTALWORDS / TOTALSENTS;
139
140  (* Report statistics:                                      *)
141  WRITELN;
142  WRITELN;
143  WRITELN( 'Text statistics:':29 );
144  WRITELN;
145  WRITELN( 'Totals      ':20 );
146  WRITELN( 'Characters      ':28, TOTALCHARS:7 );
147  WRITELN( 'Words           ':28, TOTALWORDS:7 );
148  WRITELN( 'Sentences       ':28, TOTALSENTS:7 );
149  WRITELN( 'Averages    ':20 );
150  WRITELN( 'Characters/word':28, AVGCHAR:10:2 );
151  WRITELN( 'Words/sentence ':28, AVGWORD:10:2 );
152  WRITELN( 'Maxima      ':20 );
153  WRITELN( 'Characters/word':28, LONGWORD:7 );
154  WRITELN( 'Words/sentence ':28, LONGSENT:7 );
155  WRITELN;
156  WRITELN;
157  WRITELN( '***-END OF STATISTICS-***':36 );
158  PAGE(OUTPUT)
159  END. (* TRAILTEXT *)
```

```
                    Input text:

                   THIS IS A STICKUP
   OSAGE BEACH, Mo. (UPI)--An armed bandit stuck four people to the
floor of a bait shop with an industrial-size tube of fast-drying
glue and escaped with cash and a customer's watch, police said
Saturday.
  Police said the bandit entered Wayne's Bait and Tackle Shop in
this Ozark resort community shortly after closing time Friday and
pulled a .22-caliber revolver on proprietor John Holdman.
  Holdman, his wife, Joan, their daughter, Jill, 16, and a terrified
customer were threatened by the gunman and then glued to the floor.
     (from The Chapel Hill Newspaper, July 18, 1982)


               Text statistics:

          Totals
              Characters           463
              Words                 90
              Sentences              4
          Averages
              Characters/word        5.14
              Words/sentence        22.50
          Maxima
              Characters/word       14
              Words/sentence        31


            ***-END OF STATISTICS-***
```

```
 1  PROGRAM  STATETEXT( INPUT, OUTPUT );
 2
 3  (* PURPOSE:                                                *)
 4  (*    Compute statistics about the characters, words, and *)
 5  (*    sentences in some text, using the transition         *)
 6  (*    strategy.                                            *)
 7  (*                                                         *)
 8  (* PROGRAMMER: John Potok                                  *)
 9  (*                                                         *)
10  (* INPUT:                                                  *)
11  (*    One or more lines of text to be analyzed.            *)
12  (*                                                         *)
13  (* OUTPUT:                                                 *)
14  (*    An echo of the input text.                           *)
15  (*    A summary of the text, showing how many characters, *)
16  (*    words and sentences it contains, and the largest     *)
17  (*    and average word and sentence.                       *)
18  (*                                                         *)
19  (* ASSUMPTIONS/LIMITATIONS:                                *)
20  (*    The input text is correctly punctuated.              *)
21  (*    Periods, question marks and exclamation points only *)
22  (*    occur in the last word of a sentence.                *)
23  (*    Blanks separate words.                               *)
24  (*                                                         *)
25  (* ERROR CHECKS & RESPONSES:                               *)
26  (*    None.                                                *)
27  (*                                                         *)
28  (* ALGORITHM/STRATEGY:                                     *)
29  (*    Initialize line skip flag and statistical counters. *)
30  (*    Initialize state to BLANKS.                          *)
31  (*    Read characters from text until EOF:                 *)
32  (*       Process characters according to this state table.*)
33  (*                                                         *)
34  (* STATE      CHARACTER     ACTION                NEXT STATE *)
35  (*                                                         *)
36  (* INAWORD   Blank          Echo and count word    BLANKS   *)
37  (*           End of sent.  Echo, count word and            *)
38  (*                          sentence               BLANKS   *)
39  (*           Alpha, etc.    Echo, count character  INAWORD  *)
40  (*                                                         *)
41  (* BLANKS    Blank          Echo                   BLANK    *)
42  (*           Alpha, etc.    Echo, count character  INAWORD  *)
43  (*                                                         *)
44  (*    Find the average number of characters per word and  *)
45  (*    the average number of words per sentence.            *)
46  (*    Report the statistical information.                  *)
47
```

```
48   VAR
49      CH : CHAR;                   (* One input character.        *)
50      CHARSINWORD : INTEGER;       (* Number of chars in a word.  *)
51      WORDSINSENT : INTEGER;       (* Number of words in a sent.  *)
52      TOTALCHARS: INTEGER;         (* Number of nonpunctuation    *)
53                                   (* characters in the text.     *)
54      TOTALWORDS : INTEGER;        (* Total # of words in text.   *)
55      TOTALSENTS : INTEGER;        (* Total # of sentences in text*)
56      LONGWORD : INTEGER;          (* # of chars in longest word. *)
57      LONGSENT : INTEGER;          (* # of words in longest sent. *)
58      AVGCHAR : REAL;              (* Ave. # of chars in a word.  *)
59      AVGWORD : REAL;              (* Ave.# of words in a sentence*)
60      INAWORD : BOOLEAN;           (* Current state. TRUE=INAWORD *)
61      LINESKIP : BOOLEAN;          (* Flag to remember EOLNs.     *)
62
63
64
65
66   BEGIN (* STATETEXT *)
67
68   (* Initialize counts and sums:                                 *)
69   CHARSINWORD := 0;
70   WORDSINSENT := 0;
71   TOTALCHARS:= 0;
72   TOTALWORDS := 0;
73   TOTALSENTS := 0;
74   LONGWORD := 0;
75   LONGSENT := 0;
76
77   (* Print a heading for the echo:                               *)
78   PAGE(OUTPUT);
79   WRITELN( 'Input text:':29 );
80   WRITELN;
81
82   (* Initialize starting state (allow for leading blanks):       *)
83   INAWORD := FALSE;
84
```

```
 85  (* Get and analyze the text until EOF:                         *)
 86  WHILE NOT EOF(INPUT) DO
 87     BEGIN (* Each character *)
 88
 89     (* Remember the EOLN, if any, and get the character:   *)
 90     LINESKIP := EOLN(INPUT);
 91     READ( CH );
 92
 93     (* Given the current state and character, determine the*)
 94     (* action and the new state:                            *)
 95     IF INAWORD THEN
 96        BEGIN (* State is INAWORD *)
 97        IF CH = ' ' THEN
 98           BEGIN (* Blank character *)
 99           (* This blank ends a word.  Echo it:              *)
100           WRITE( CH );
101           WORDSINSENT := WORDSINSENT + 1;
102           TOTALCHARS := TOTALCHARS+ CHARSINWORD;
103           (* Check for the longest word:                    *)
104           IF CHARSINWORD > LONGWORD THEN
105               LONGWORD := CHARSINWORD;
106           (* Reset the character counter:                   *)
107           CHARSINWORD := 0;
108           (* Change states:                                 *)
109           INAWORD := FALSE
110           END (* Blank character *)
111        ELSE IF CH IN [ '.', '!', '?' ] THEN
112           BEGIN (* End of sentence *)
113           (* This char ends a word and sentence. Echo it:  *)
114           WRITE( CH );
115           WORDSINSENT := WORDSINSENT + 1;
116           TOTALWORDS := TOTALWORDS + WORDSINSENT;
117           TOTALSENTS := TOTALSENTS + 1;
118           TOTALCHARS := TOTALCHARS + CHARSINWORD;
119           (* See if this is the longest sentence:           *)
120           IF WORDSINSENT > LONGSENT THEN
121              LONGSENT := WORDSINSENT;
122           (* See if this is the longest word:               *)
123           IF CHARSINWORD > LONGWORD THEN
124              LONGWORD := CHARSINWORD;
125           (* Reset the word and character counter:          *)
126           WORDSINSENT := 0;
127           CHARSINWORD := 0;
128           (* Change states:                                 *)
129           INAWORD := FALSE
130           END (* End of sentence *)
131        ELSE
132           BEGIN (* Other nonblank *)
133           (* Echo and count it (if not punctuation):        *)
134           WRITE( CH );
135           IF NOT (CH IN [ ',', ':', ';' ]) THEN
136              CHARSINWORD := CHARSINWORD + 1
137           END (* Other nonblank *)
138        END (* State is INAWORD *)
```

```
139
140      ELSE
141         BEGIN (* State is BLANKS *)
142         IF CH = ' ' THEN
143            WRITE( CH )
144         ELSE
145            BEGIN (* A nonblank *)
146            (* This character begins a new word.  Echo it:    *)
147            WRITE( CH );
148            CHARSINWORD := CHARSINWORD + 1;
149            (* Change states:                                 *)
150            INAWORD := TRUE
151            END; (* A nonblank *)
152         END; (* State is BLANKS *)
153
154      (* If we had an EOLN, then begin a new output line:     *)
155      IF LINESKIP THEN
156         WRITELN
157
158      END; (* Each character *)
159
160   (* Compute the average number of characters per word and  *)
161   (* average number of words per sentence:                   *)
162   AVGCHAR := TOTALCHARS / TOTALWORDS;
163   AVGWORD := TOTALWORDS / TOTALSENTS;
164
165   (* Report statistics:                                      *)
166   WRITELN;
167   WRITELN;
168   WRITELN( 'Text statistics:':30 );
169   WRITELN;
170   WRITELN( 'Totals      ':20 );
171   WRITELN( 'Characters      ':28, TOTALCHARS:7 );
172   WRITELN( 'Words           ':28, TOTALWORDS:7 );
173   WRITELN( 'Sentences       ':28, TOTALSENTS:7 );
174   WRITELN( 'Averages    ':20 );
175   WRITELN( 'Characters/word':28, AVGCHAR:10:2 );
176   WRITELN( 'Words/sentence ':28, AVGWORD:10:2 );
177   WRITELN( 'Maxima      ':20 );
178   WRITELN( 'Characters/word':28, LONGWORD:7 );
179   WRITELN( 'Words/sentence ':28, LONGSENT:7 );
180   WRITELN;
181   WRITELN;
182   WRITELN( '***-END OF STATISTICS-***':36 );
183   PAGE(OUTPUT)
184   END. (* STATETEXT *)
```

```
                    Input text:

                   THIS IS A STICKUP

  OSAGE BEACH, Mo. (UPI)--An armed bandit stuck four people to the
floor of a bait shop with an industrial-size tube of fast-drying
glue and escaped with cash and a customer's watch, police said
Saturday.
  Police said the bandit entered Wayne's Bait and Tackle Shop in
this Ozark resort community shortly after closing time Friday and
pulled a .22-caliber revolver on proprietor John Holdman.
  Holdman, his wife, Joan, their daughter, Jill, 16, and a terrified
customer were threatened by the gunman and then glued to the floor.
     (from The Chapel Hill Newspaper, July 18, 1982)


                Text statistics:

         Totals
             Characters          479
             Words                90
             Sentences             4
         Averages
             Characters/word       5.32
             Words/sentence       22.50
         Maxima
             Characters/word      15
             Words/sentence       31


           ***-END OF STATISTICS-***
```

PROGRAM EXERCISES

1. In TRAILTEXT, make sure that the segment at lines 90-92 is properly stated and positioned for its intended purpose. If not, what changes should be made?

2. (a) Why does TRAILTEXT say that the longest word has 14 characters, while STATETEXT says 15?
   (b) What are all the extra characters counted by STATETEXT?
   (c) Discuss why two programs that should do the "same thing" really do not, and what can be done to prevent such an outcome.

3. Draw the state transition diagram for STATETEXT.

4. Change the transition diagram for STATETEXT so that hyphenated words are counted as separate words, and quotations and contractions are handled properly.

5. Change STATETEXT to conform to the new transition diagram created in the previous exercise.

Part III: Array Algorithms

INTRODUCTION

Array algorithms are perhaps the most important group of algorithms in programming. This is no doubt because arrays have been available in one form or another throughout the history of programming. A large body of experience with arrays has by now been accumulated.

Arrays are usually used to represent lists of information. Algorithms to manipulate these lists are the emphasis of this part. Arrays have traditionally been used to represent other forms of data structures as well. That idea will be reviewed briefly at the end of this part. In addition, character strings are arrays with some special properties; they are the subject of Part V.

Because this part uses many declarations, they are presented once—here. Only a few more will be introduced throughout the discussion.

Assume that we have these declarations:

```
CONST
   MAX = ...;               (* Maximum array index (integer). *)
   MAXPLUS1 = ...;          (* One more than MAX.             *)

TYPE
   INDEX = 1..MAX;          (* List subscript type.           *)
   COUNT = 0..MAX;          (* Actual number of elements.     *)
   ITEM = ...;              (* Assume a scalar type for now.  *)
   LIST = ARRAY[INDEX] OF ITEM;
                            (* A generalized list type.       *)

VAR
   A : LIST;                (* A list of items.               *)
   N : COUNT;               (* Current # of list elements.    *)
   X : ITEM;                (* Any possible list element.     *)
   I : INDEX;               (* List index used in loops.      *)
   POSN : 0..MAXPLUS1;      (* A possible list position, or   *)
                            (* a position just outside it.    *)
```

A lower bound of 1 is used because it is easier to visualize arrays with subscript types of 1..MAX, and because the current number of elements, N, can also represent the largest currently used array position.

Let us assume that there are N items in list A now, that A can contain MAX items, that X can contain a value from or for A, and that A can be indexed by I and POSN. We have:

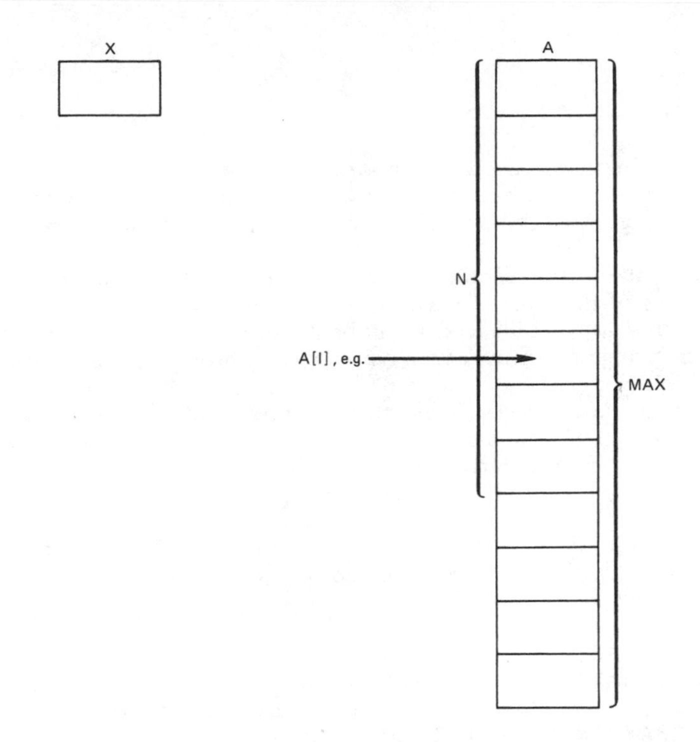

A list is considered ''empty'' when it contains no items—that is, when the number of items (the number of positions being used to hold information) is 0. Thus, the list A can be initialized to empty by setting N to zero. (Obviously, the array holding the conceptual list still exists.) We can see if the list is empty by checking for N equal to 0, and know that it is full when N is equal to MAX.

INPUT AND OUTPUT

There are three input loop control methods for simple values, as you saw in Part I. These same methods can be applied to array input—with a few modifications.

Assume that the input is arranged with one value (list element) on each line. The most straightforward algorithm assumes that the number of elements appears as an extra first data value:

```
(* Get N, then read N elements into the list A:              *)
READLN( N );
FOR I:=1 TO N DO
   READLN( A[I] )
```

A simple algorithm that is used quite often merely appends a value to the "end" (next unused location) of the list:

```
(* Append the value of X to the list A of size N:            *)
N := N + 1;
A[N] := X
```

Now, given an appropriate constant called SENTINEL, the sentinel input method is:

```
(* Read values into A until the SENTINEL is found:            *)
N := 0;
READLN( X );
WHILE X <> SENTINEL DO
   BEGIN (* each value *)

   (* Append the value of X to A:                             *)
   N := N + 1;
   A[N] := X;

   READLN( X )
   END (* each value *)
(* N is now the number of items in A.                         *)
```

The EOF method is somewhat similar to both of the other methods:

```
(* Read values into the list A until EOF:                       *)
N := 0;
WHILE NOT EOF(INPUT) DO
   BEGIN (* each value *)

   (* Count the new element and get it directly:                *)
   N := N + 1;
   READLN( A[N] )

   END (* each value *)
(* N is now the number of items in A.                           *)
```

Ex. 1: Rewrite the three input algorithms so that they also echo each value with its index in the list. Include column headings.

It is not possible to append a value to an array that is already full:

```
(* Append the value of X to the list A, if possible:            *)
IF N < MAX THEN
   BEGIN (* append *)
   N := N + 1;
   A[N] := X
   END (* append *)
```

So a safer sentinel method is:

```
(* Read values into list A until SENTINEL is encountered, *)
(* or until the list is full:                             *)
N := 0;
READLN( X );
WHILE X <> SENTINEL DO
   BEGIN (* each value *)

   (* Append X to A if possible:                          *)
   IF N < MAX THEN
      BEGIN (* append *)
      N := N + 1;
      A[N] := X
      END (*  append *)
   ELSE
      WRITELN( 'Excess data:', X );

   READLN( X )
   END (* each value *)
```

An algorithm that protects itself against errors as this one does is said to be "robust." Robustness is a desirable property because it keeps the program in control even under adverse conditions and warns the program user of actual or potential problems. A basic algorithm also becomes larger and more complicated when it is made robust, as we can see by comparing the two versions of the sentinel input method. For this reason, only the basic versions of algorithms are presented in this book. However, many of the exercises ask you to think about robustness.

Ex. 2: Rewrite the EOF algorithm to avoid list overflow (where N becomes greater than MAX), and to display any excess data.

†Ex. 3: Rewrite the EOF algorithm to read MAX values or until EOF, whichever comes first.

The last topic to be presented in this section is output. It is quite straightforward. A typical form would include a title and column headings:

```
(* Print the contents of list A:                              *)
WRITELN( 'List of items:' );
WRITELN;
WRITELN( 'Index', 'Content':9 );
FOR I:=1 TO N DO
   WRITELN( I:5, A[I]:9 )
```

Ex. 4: Assume that the items in A are integers of type 0..100. Rewrite the basic output algorithm to include a histogram (a row of asterisks representing each quantity).

BASIC CALCULATIONS

This brief section shows three general kinds of calculations that are done with the *numeric* contents of arrays.

Assume for the moment that we have these variables in addition to those shown in the Introduction:

```
SUM : ITEM;            (* The sum of list elements.        *)
FACTOR : ITEM;         (* Represents any value.            *)
B : LIST;              (* Two lists just like A.           *)
C : LIST;
```

Part I shows how to find the sum of input values. We can also sum the elements of the list A:

```
(* Find the sum of the N items in A:  (N could be 0)         *)
SUM := 0;
FOR I:=1 TO N DO
   SUM := SUM + A[I]
```

It is also common to perform some calculation on each value in the list:

```
(* Multiply each item in A by FACTOR:                        *)
FOR I:=1 TO N DO
   A[I] := A[I] * FACTOR
```

Ex. 5: Write an algorithm to print the list A, showing the percent of the SUM (as above) that is represented by each item.

If there are N items in both A and B, the corresponding items in each list can be added to form a list of sums, C:

```
(* Put the element-wise sum of A and B into C:               *)
FOR I:=1 TO N DO
   C[I] := A[I] + B[I]
```

Most calculations done on the values of items in a list are variations of these three algorithms.

†*Ex. 6: Write an algorithm to find and print the items in A, their running total, and their values increased by 4%.*

SEARCHES

A list is a collection of different but conceptually related values. The preceding sections presented algorithms that do not differentiate among the items in a list. It is also often necessary to search through a list, looking for particular values of interest. This section presents the most common of these searching algorithms.

Assume that we have this variable:

```
LARGE : ITEM;          (* The largest value in list A.   *)
```

Many of the algorithms that manipulate a sequence of input values have counterpart array algorithms. This is because values can be taken from input *or* from an array and used in one of the more fundamental algorithms. For example, the largest item in a non-empty list can be found:

```
(* Search A for the largest of its N values:                 *)
LARGE := A[1];            (* The largest so far.             *)
FOR I:=2 TO N DO
   IF A[I] > LARGE THEN
      LARGE := A[I]
```

Ex. 7: Rewrite this algorithm to find the largest value and its position, POSN, in the list.

Similarly, all values less than X can be selected and printed:

```
(* Print all values in A that are less than X:               *)
FOR I:=1 TO N DO
   IF A[I] < X THEN
      WRITELN( A[I] )
```

Ex. 8: Rewrite this algorithm to count and sum the values found and to print them with their indices.

Assume for now that the list is "unordered." That is, the values in the list are not in any particular order, such as alphabetic order or numeric order.

To find out if and where a given value is in the list, we must look at each element in turn until the value is found or until the entire list has been checked. This variable is necessary:

```
    FOUND : BOOLEAN;        (* TRUE if we found the value.    *)
```

The algorithm looks like this:

```
(* Find the position, POSN, of the value of X in list A:   *)
POSN := 1;
FOUND := FALSE;
WHILE (POSN <= N) AND NOT FOUND DO
   IF A[POSN] = X THEN
      FOUND := TRUE
   ELSE
      POSN := POSN + 1
(* If FOUND is TRUE, then POSN now points to the value.    *)
```

†*Ex. 9: What is the value of POSN if FOUND ends up FALSE?*

†*Ex. 10: How does this algorithm change if the list is to be searched backward? What final values might POSN have?*

This algorithm is the basic *linear search,* a search that tests each item in the sequence until the proper value is found. There are many variations of the basic linear search. A very clear one promoted by L. V. Atkinson (see the Bibliography) uses the variable:

```
    STATUS : (SEARCHING, (* Current status of the search.  *)
              MISSING,
              FOUNDIT);
```

The algorithm is then:

```
(* Find the position, POSN, of the value of X in list A:   *)
POSN := 1;
STATUS := SEARCHING;
REPEAT
   IF POSN > N THEN
      STATUS := MISSING
   ELSE IF A[POSN] = X THEN
      STATUS := FOUNDIT
   ELSE
      POSN := POSN + 1
UNTIL STATUS <> SEARCHING
(* If STATUS = FOUNDIT, then POSN is its position.         *)
```

The searches shown so far were set up for unordered lists. More frequently—and more interestingly—lists are *ordered.* That is, the values they contain are arranged in alphabetic or numeric order. If the values increase from the beginning of the list toward the end, the list is in ascending order; if they decrease, it is in descending order.

If a list is in ascending order, the basic linear search for a given value will still work. More frequently, we want to know the position in the list where a given value *should be,* whether or not it is there now:

```
(* Find the position for X in the ordered list A:          *)
POSN := 1;
FOUND := FALSE;
WHILE (POSN <= N) AND NOT FOUND DO
   IF A[POSN] >= X THEN
      FOUND := TRUE
   ELSE
      POSN := POSN + 1
(* If FOUND, then POSN is where X should be within A.       *)
```

†*Ex. 11: Write the additional statements necessary to report the success of actually having found X in the list.*

Ex. 12: Rewrite the algorithm to search the list backward, assuming that it is in ascending order.

This search can be simplified by using the value of X as a sentinel at the end of the list, ensuring that a position for X will *always* be found:

```
(* Find the position for X in A, using X as a sentinel:    *)
A[N+1] := X;
POSN := 1;
WHILE A[POSN] < X DO
   POSN := POSN + 1
(* If POSN <= N, then it shows where X is or should be.     *)
```

†*Ex. 13: Does this algorithm work for all situations—that is, for an empty list and a completely full list? Does the previous algorithm work in all situations?*

Linear searches are easy to write and to understand, and so they are quite useful and common. Their use is usually restricted to short lists, however, because they are not "efficient" enough for long lists.

Discussions of algorithms often speak of efficiency. One algorithm is *more efficient or faster* than another if it performs the same task using fewer comparisons or assignments of information. Efficiency is a desirable property of algorithms, such as searches, that may process large amounts of information, because computer processing time costs money.

There is a faster way to search an ordered list, called a *quadratic search.* We will need these variables:

```
STEP : INDEX;            (* Interval between items tested. *)
FOUND : BOOLEAN          (* TRUE if given value was found. *)
```

A successful search can be illustrated as a two part algorithm this way:

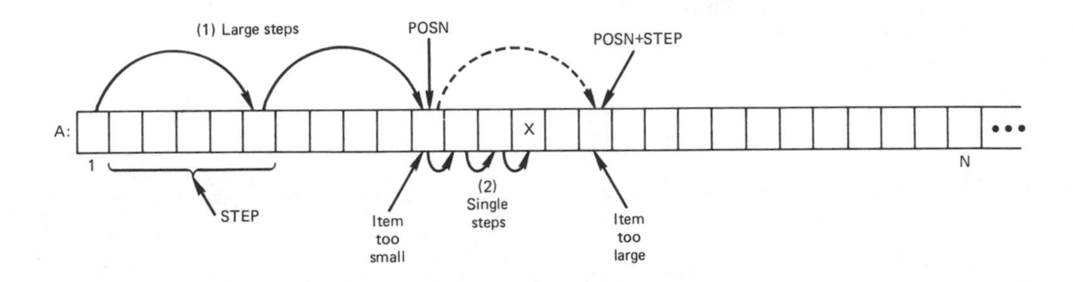

The algorithm is:

```
(* Quadratic search to find the position for X in A:           *)

(* Take large steps through A to get near X's position:        *)
STEP := ROUND( SQRT(N+1) );   (* Selects an optimum value. *)
POSN := 1;
FOUND := FALSE;
WHILE (POSN+STEP <= N) AND NOT FOUND DO
   IF A[POSN+STEP] > X THEN
      FOUND := TRUE
   ELSE
      POSN := POSN + STEP;

(* Now POSN is close.  Single-step to the exact location: *)
(* (A standard linear search.)                             *)
FOUND := FALSE;
WHILE (POSN <= N) AND NOT FOUND DO
   IF A[POSN] >= X THEN
      FOUND := TRUE
   ELSE
      POSN := POSN + 1
```

†*Ex. 14: Does the search work for an empty list, a one-item list, and when all items in the list are less than X?*

There is an even faster search for ordered lists called a *binary search.* It requires these variables (as well as FOUND):

```
LOW : 0..MAXPLUS1;    (* Low index of a search interval.*)
HIGH : 0..MAXPLUS1;   (* High index of the interval.    *)
```

A typical successful search can be illustrated like this:

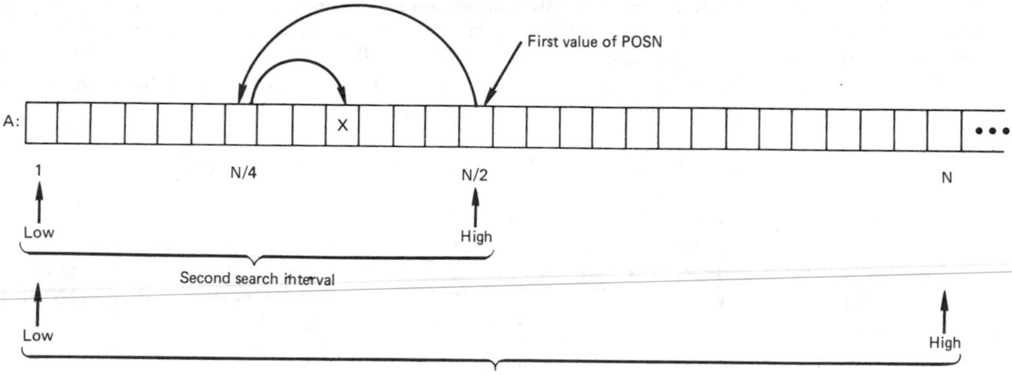

A typical implementation would be:

```
(* Binary search to find the position for X in A:               *)

(* Set bounds for the 1st search interval (entire list):        *)
LOW := 1;
HIGH := N;
(* Narrow the search by cutting the interval in half            *)
(* until the position for X is found:                           *)
FOUND := FALSE;
WHILE (LOW <= HIGH) AND NOT FOUND DO
   BEGIN (* each interval *)

   (* Find the middle of the interval:                          *)
   POSN := (LOW + HIGH) DIV 2;

   (* See if X would lie before this position, after it,        *)
   (* or right at this position:                                *)
   IF X < A[POSN] THEN
      (* Exclude the upper half of the interval:                *)
      HIGH := POSN - 1
   ELSE IF X > A[POSN] THEN
      (* Exclude the lower half:                                *)
      LOW := POSN + 1
   ELSE
      FOUND := TRUE

   END (* each interval *)
(* If FOUND, then POSN tells where.  Otherwise, LOW tells *)
(* where X should be.                                         *)
```

†Ex. 15: What is the value of LOW when X is larger than all values in the list? When X is smaller than all values?

Ex. 16: Does this search work when A is an empty list or a one-item list?

Ex. 17: Why will quadratic and binary searches not work with unordered lists?

Unordered lists must be searched with linear searches. The linear searches are also fast enough to use for short ordered lists containing up to a couple of dozen elements, unless the lists are searched very often. The quadratic search is actually unnecessary; the binary search is faster. It is included here because it is conceptually related to other algorithms, as will be seen. The binary search should be used most of the time.

Ex. 18: Is it possible to use a binary search on an array whose index type is CHAR? If its index type is enumerated?

The search algorithms in this section can be implemented in many different ways, but the reason for searching an ordered list remains the same: to find where a given value is in the list or, if not there, where it *should* be. Knowing where it should be is useful if we intend to put that value into the list. That is the subject of the next section.

SORTS

A collection of related information is most often stored in an ordered list. This is because information can be located more quickly in ordered lists, as shown in the previous section. The information may originally be gathered at random, however, so we need algorithms to put it in order. This is called ordering or, more often, *sorting*. This section presents several of the many algorithms for sorting lists.

Recall the shift algorithm from Part I. We can also shift the items in a list, moving them one position toward the end. If we shift all items whose positions are POSN or greater, a "gap" (an unused location) will be left in the list at POSN:

```
(* Shift all items in the sublist from POSN to N one step *)
(* toward the end of list A, assuming that A is not full: *)
FOR I:=N DOWNTO POSN DO
   A[I+1] := A[I];
(* Adjust N to the new size of A:                          *)
N := N + 1
```

†*Ex. 19: What would be the effect if the loop iterated from POSN to N?*

Ex. 20: Rewrite the algorithm so that it shifts the same sublist of A a given number of positions.

The value of X can now be assigned into the gap left at POSN:

```
(* Put the value of X at POSN:                             *)
A[POSN] := X
```

If the shift and the assignment of X are combined, they form an insertion algorithm:

```
(* Insert the value of X into list A at position POSN:    *)
FOR I:=N DOWNTO POSN DO
   A[I+1] := A[I];
N := N + 1;
A[POSN] := X
```

†*Ex. 21: Does the algorithm work if POSN is one greater than N? If N is 0?*

Ex. 22: Given that B is a list containing M items, write an algorithm to append all of B to A.

Ex. 23: Given that B is a list of M items, write an algorithm to insert all of B into A starting at POSN.

Recall the searching algorithms from the previous section. Given an ordered list, these algorithms set POSN to the position where the value of X should be in the list. We know how to insert X at that position. If we do so, A is still ordered.

Now several algorithms can be combined to build a list in ascending order directly from unordered input:

```
(* Build the ordered list A from unordered input:              *)
N := 0;
WHILE NOT EOF(INPUT) DO
   BEGIN (* each input value *)

   (* Get and echo this value:                                 *)
   READLN( X );
   WRITELN( X );

   (*Find the position for X in A, using X as a sentinel: *)
   A[N+1] := X;
   POSN := 1;
   WHILE A[POSN] < X DO
      POSN := POSN + 1;

   (* Append or insert X, given the result of the search: *)
   IF POSN > N THEN
      (* X belongs where we put it as a sentinel!              *)
      N := N + 1
   ELSE
      (* X belongs at position POSN, so insert it:             *)
      BEGIN (* insert *)
      FOR I:=N DOWNTO POSN DO
         A[I+1] := A[I];
      N := N + 1;
      A[POSN] := X
      END (* insert *)

   END (* each input value *)
```

Ex. 24: Show how to protect this algorithm from too much data.

†*Ex. 25: Simplify the "append or insert" portion of the algorithm into one segment that is always executed. (That is, eliminate the IF statement.)*

Ex. 26: Show how to change the algorithm so that it builds the ordered list A from the M values stored in an unordered list B.

If the list will become very long, a binary search should be used instead of a linear search.

Ex. 27: Rewrite the algorithm so that it searches backward through the list, shifting at the same time so that the new value can be assigned immediately.

Recall the algorithm to exchange or swap the values of two simple variables. We can do the same with two items in a list. Another index variable is necessary:

```
    J : INDEX;              (* List index used in loops, etc. *)
```

The algorithm (assuming that I and J already have values) is simply:

```
(* Swap the values of A[I] and A[J]:                           *)
X := A[I];
A[I] := A[J];
A[J] := X
```

We now have all the algorithms that we need to write an *insertion sort.* An insertion sort is one of many sorting algorithms that take an unordered list and put its elements in order—without using another list:

```
(* Insertion sort to put A's elements in ascending order: *)

(* Take each item in A, search backward in the list for   *)
(* its proper position, and reinsert it at that point:     *)
FOR I:=2 TO N DO
   BEGIN (* each item *)

   (* Isolate the item, using X:                           *)
   X := A[I];

   (* Search backward (from this position) for X's proper *)
   (* position:                                            *)
   POSN := I - 1;
   FOUND := FALSE;
   WHILE (POSN >= 1) AND NOT FOUND DO
      IF X < A[POSN] THEN
         POSN := POSN - 1
      ELSE
         FOUND := TRUE;

   (* Reinsert the value into the list at POSN:            *)
   (* (Note that 1 <= POSN+1 <= I. )                       *)
   FOR J :=I-1 DOWNTO POSN+1 DO
      A[J+1] := A[J];
   A[POSN+1] := X

   END (* each item *)
```

†*Ex. 28: Why must J iterate only within the limits shown?*

Ex. 29: Could a binary search be substituted for the backward linear search?

Ex. 30: Could a forward linear search be substituted for the backward one?

Remember that the shift algorithm within the insertion steps backward through the list. The backward search and the shift can thus be combined into one loop:

```
(* Insertion sort to put A's elements in ascending order: *)

(* Take each item in A, search backward for its proper    *)
(* position, and reinsert it:                             *)
FOR I:=2 TO N DO
    BEGIN (* each item *)

    (* Isolate the item:                                   *)
    X := A[I];

    (* Search backward (from this position) for X's        *)
    (* position, making room for X (shifting) as we go:    *)
    POSN := I - 1;
    FOUND := FALSE;
    WHILE (POSN >= 1) AND NOT FOUND DO
        BEGIN (* search & shift *)
        IF X < A[POSN] THEN
            BEGIN (* shift *)
            A[POSN+1] := A[POSN];
            POSN := POSN - 1
            END (* shift *)
        ELSE
            FOUND := TRUE
        END; (* search & shift *)

    (* Reinsert the value:                                 *)
    A[POSN+1] := X

    END (* each item *)
```

This combined search and shift makes the insertion sort faster.

There is another common sorting algorithm called a *selection sort:*

```
(* Selection sort to put A into ascending order:              *)

(* For each successively shorter sublist of A, put its        *)
(* smallest element in its first position:                    *)
FOR I:=1 TO N-1 DO
   BEGIN (* each sublist *)

   (* Select the smallest item in the sublist from A[I]       *)
   (* to A[N]:   (Use X to hold the smallest.)                *)
   X := A[I];
   POSN := I;
   FOR J:=I+1 TO N DO
      IF A[J] < X THEN
         BEGIN (* save *)
         X := A[J];
         POSN := J
         END; (* save *)

   (* Exchange the smallest item with the first item in       *)
   (* the sublist:                                            *)
   A[POSN] := A[I];
   A[I] := X

   END (* each sublist *)
```

The last sorting algorithm to be shown here is based upon a simpler algorithm that moves the largest item in a list to the last position (in several steps) and moves the smallest item one step toward the front:

```
(* "Bubble" the smallest item up one step.                    *)
(* Compare and order each successive pair of items in A:      *)
FOR J:= 1 TO N-1 DO
   BEGIN (* each pair *)

   (* Put this pair into ascending order:                     *)
   IF A[J] > A[J+1] THEN
      BEGIN (* swap *)
      X := A[J];
      A[J] := A[J+1];
      A[J+1] := X
      END (* swap *)

   END (* each pair *)
```

Since the largest item is now at the end of the list, it need not be compared again, but the algorithm should be repeated to make the smaller values "bubble" toward the top. It should stop when the list is in order.

Assume that we have these variables:

```
INORDER : BOOLEAN;     (* TRUE if the list is ordered.      *)
LIMIT : COUNT;         (* Extent of the unordered part.     *)
```

The algorithm, called a *bubble sort,* is:

```
(* Bubble sort to put A into ascending order:                 *)
LIMIT := N;
REPEAT

   (* Assume that the list is already ordered:                 *)
   INORDER := TRUE;

   (* Limit the comparisons to the next shorter sublist:      *)
   LIMIT := LIMIT - 1;

   (* Compare and order successive pairs of list items:       *)
   FOR J:=1 TO LIMIT DO
      BEGIN (* each pair *)

      (* Put this pair in ascending order:                     *)
      IF A[J] > A[J+1] THEN
         BEGIN (* swap *)
         (* The list is not yet in order:                      *)
         INORDER := FALSE
         X := A[J];
         A[J] := A[J+1];
         A[J+1] := X;
         END (* swap *)

      END (* each pair *)

UNTIL INORDER
```

Efficiency can be an important consideration when choosing sorting algorithms. The bubble sort is a good sort for lists that are nearly ordered already, because it will stop as soon as possible. The selection sort is much better for longer, randomly ordered lists. The insertion sorts are a little better still. There are several much faster—and more complicated—sorts available; see the books about algorithms in the Bibliography.

OTHER OPERATIONS ON LISTS

The sections on searches and sorts also introduced some simpler algorithms to append the entire contents of one list to another, and to insert an entire list into the middle of another. Such algorithms can be quite useful. In fact, there are enough of these general operations on lists to devote this section to them.

These variables will be used quite often:

```
B : LIST;                (* A list similar to A.          *)
M : COUNT;               (* Actual number of items in B.  *)
APOSN : INTEGER;         (* A loop index for A.           *)
BPOSN : INTEGER;         (* A loop index for B.           *)
```

It is often necessary to delete unwanted items from an ordered or unordered list. Suppose that, as an example, all even values are to be removed, while the odd ones are to be kept:

```
(* Delete all unwanted values from A:                          *)
POSN := 0;
FOR I:=1 TO N DO
   BEGIN (* each item *)

   (* Save only values that are ok: (odd, for example)        *)
   IF ODD( A[I] ) THEN
      BEGIN (* save it *)
      POSN := POSN + 1;
      A[POSN] := A[I]
      END (* save it *)
   END; (* each item *)
(* Reset N to the actual number of items left in A:            *)
N := POSN
```

The algorithm is similar to an append; it packs the "good" items into the front end of the list.

†*Ex. 31: Write an algorithm to delete from the ordered list A all items whose values are less than some constant called LIMIT. Write an algorithm to delete all items larger than LIMIT.*

Ex. 32: Rewrite the deletion algorithm so that each unwanted item in an unordered list is replaced by the last (at that moment) item in the list, thus also shortening the list.

If we have two separate, similarly ordered lists A and B, containing N and M items, respectively, we often want to see the contents of the two lists *merged* into one long ordered list:

```
(* Merge lists A and B into one output column:                    *)

(* Print items from both lists in ascending order:                *)
APOSN := 1;
BPOSN := 1;
WHILE (APOSN <= N) AND (BPOSN <= M) DO
   BEGIN (* compare items *)

   (* Select the smaller of the two current items:                *)
   IF A[APOSN] < B[BPOSN] THEN
      BEGIN (* print A's item *)
      WRITELN( A[APOSN] );
      APOSN := APOSN + 1
      END (* print A's item *)
   ELSE
      BEGIN (* print B's *)
      WRITELN( B[BPOSN] );
      BPOSN := BPOSN + 1
      END (* print B's *)

   END; (* compare items *)

(* Print items remaining in A (if any):                           *)
WHILE APOSN <= N DO
   BEGIN (* each item *)
   WRITELN( A[APOSN] );
   APOSN := APOSN + 1
   END; (* each item *)

(* Print items remaining in B (implies none left in A):           *)
WHILE BPOSN <= M DO
   BEGIN (* each item *)
   WRITELN( B[BPOSN] );
   BPOSN := BPOSN + 1
   END (* each item *)
```

†*Ex. 33: Write an algorithm to merge A and B into a third list C, using K to count its items.*

Ex. 34: Write an algorithm to merge the M items of B directly into A, so that A contains all N plus M items in order. (The shortest algorithm steps backward through the two ordered lists.)

Rather than merge two lists, we often want to split an ordered or unordered list into two lists, basing the split on some property of the list elements. For example, using odd versus even again:

```
(* Split A into lists of even and odd items, putting the   *)
(* even items into B and keeping the odd ones in A:         *)
POSN := 0;
M := 0;
FOR I:=1 TO N DO
   BEGIN (* each item *)

   (* Determine which list it belongs in and put it there *)
   IF ODD( A[I] ) THEN
      BEGIN (* save in A *)
      POSN := POSN + 1;
      A[POSN] := A[I]
      END (* save in A *)
   ELSE
      BEGIN (* append to B *)
      M := M + 1;
      B[M] := A[I]
      END (* append to B *)

   END; (* each item *)
(* Reset N to the new size of A:                            *)
N := POSN
```

†*Ex. 35: Write an algorithm to split the ordered list A into two parts, keeping items whose values are larger than LIMIT in A, while putting the rest into B.*

When an ordered list contains duplicate entries, perhaps as the result of a merge, we may want to "compress" the list—that is, keep one copy of each *different* value:

```
(* Compress the ordered list A (remove duplicate values): *)
POSN := 1;
FOR I:=2 TO N DO
   BEGIN (* each item *)

   (* Save this item only if different from previous one: *)
   IF A[I] <> A[I-1] THEN
      BEGIN (* save *)
      POSN := POSN + 1;
      A[POSN] := A[I]
      END (* save *)
   END; (* each item *)
(* Reset N to the new number of elements (if any):         *)
IF N > 0 THEN
   N := POSN
```

Ex. 36: Rewrite this algorithm so that it will not assign an item to itself when there is a sequence of different items.

Ex. 37: If you wished to remove duplicate items from an unordered list, which of these algorithms would be the most efficient?
(a) Sort it, then remove duplicates as before.
(b) Step through the list, comparing each item to all the items saved so far, and save it if it is different.
(c) As in (b), but when saving each new item, insert it among the other saved items in proper order, so that the more efficient binary search can be used.
The answer could be found using a mathematical analysis of each algorithm, which is beyond the scope of this book, but you could also find the answer by experiment.

Finally, we may want to compare two lists for equality. This variable can be used:

```
   EQUAL : BOOLEAN;       (* TRUE if two lists are equal.   *)
```

The algorithm would be:

```
(* Compare lists A and B (both length N) for equality:     *)
EQUAL := TRUE;
POSN := 1;

(* Compare corresponding items in the two lists:           *)
WHILE (POSN <= N) AND EQUAL DO
   BEGIN (* corresponding items *)

   IF A[POSN] <> B[POSN] THEN
      EQUAL := FALSE
   ELSE
      POSN := POSN + 1

   END (* corresponding items *)
(* A and B are equal if EQUAL is still true.               *)
```

†*Ex. 38: How would you change this algorithm to compare two lists of possibly unequal lengths?*

ARRAYS USED AS FUNCTIONS

Arrays were shown to be important data structures for holding lists of information. Another important use is as "functions" that return tabulated rather than calculated values. A function accepts one value and returns another value related somehow to the first.

As a specific example, assume that we have these declarations:

```
TYPE
   MONTHS =                  (* The months of the year.          *)
      (JAN, FEB, MAR, APR, MAY, JUN,
       JUL, AUG, SEP, OCT, NOV, DEC);

VAR
                             (* The # of days in each month:     *)
   DAYSIN : ARRAY[MONTHS] OF 1..31;
   MONTH : MONTHS;           (* Index into DAYSIN.               *)
```

Then DAYSIN can be used as a function to convert a value of type MONTHS into the number of days in that month. It could be initialized this way:

```
(* Initialize DAYSIN to hold the # of days in each month: *)
FOR MONTH:=JAN TO DEC DO
   IF MONTH IN [JAN,MAR,MAY,JUL,AUG,OCT,DEC] THEN
      DAYSIN[MONTH] := 31
   ELSE IF MONTH IN [APR,JUN,SEP,NOV] THEN
      DAYSIN[MONTH] := 30
   ELSE (* FEB *)
      DAYSIN[MONTH] := 28       (* Not a leap year.          *)
```

Given any month in MONTH, then DAYSIN[MONTH] is the correct number of days.

An array used as a function would never be sorted, nor would values be inserted, deleted, appended, and so on, after the array has been initialized. This is because the value at each location is related directly to its index value, and does not change.

In general, any finite scalar (character, boolean, enumerated type, or subrange) can be mapped onto any other type.

As another example, we can set up a function to map characters into their corresponding character classes, which we found useful in Part II.

Assume these declarations:

```
TYPE
   CLASSES =                   (* Character processing classes:  *)
      (BLANK,                  (* The blank only.                *)
       PUNCT,                  (* All end-of-word punctuation.   *)
       ALPHA,                  (* Upper- and lowercase letters.  *)
       DIGIT,                  (* The character digits.          *)
       OTHER);                 (* All other characters.          *)

VAR
                               (* CHAR to CLASSES "function":    *)
   CLASSOF : ARRAY[CHAR] OF CLASSES;
   CH : CHAR;                  (* Index into CLASSOF.            *)
```

The array can be initialized this way:

```
(* Initialize the CLASSOF array to be used as a function: *)

(* Let the default class value be OTHER:                 *)
FOR CH:=CHR(0) TO CHR(255) DO
   CLASSOF[CH] := OTHER;

(* Now set the classes for specific characters:          *)
FOR CH:='A' TO 'Z' DO
   BEGIN (* each letter *)
   (* Set uppercase:                                     *)
   CLASSOF[CH] := ALPHA;
   (* Set lowercase:                                     *)
   CLASSOF[ CHR(ORD(CH)-64) ] := ALPHA (* add 32 in ASCII *)
   END; (* each letter *)
FOR CH:='0' TO '9' DO
   CLASSOF[CH] := DIGIT;
FOR CH:='.' TO '?' DO          (* use '!' to '?' in ASCII  *)
   IF CH IN ['.', ',', ':', ';', '!', '?'] THEN
      CLASSOF[CH] := PUNCT;
CLASSOF[' '] := BLANK
```

Now, given CH, its class is found using CLASSOF[CH]. The GETNEXT procedure shown in PART II could have an array like this available as a global variable.

The fact that we *can* use an array as a function is not adequate reason to do so, but there are two good reasons: the array "function" is more efficient than a genuine function, because the value need not be recalculated each time, and, whereas genuine functions can only return single-valued results, arrays can hold structured values as well.

ARRAYS USED AS OTHER STRUCTURES

Arrays have traditionally been used to simulate data structures other than simple lists—stacks, queues, trees, and many others. The tradition arose in the early days of programming and persists among users of the early programming languages. Modern languages have pointers; these are more appropriate for creating new data structures, as is shown in Part VII. However, some of the traditional algorithms are still useful—those dealing with stacks. (A stack is defined as a list that is accessed from only one end, called its "top," like a stack of plates.)

Assume that we have these declarations:

```
TYPE
    STACK =                     (* A stack of items.            *)
        RECORD
        VAL : LIST;             (* The stack contents.          *)
        TOP : COUNT             (* Number of items in VAL.      *)
        END;
VAR
    STK : STACK;
```

Assume that the last item in the list (the item with the largest subscript) is the top of the stack. TOP always points to it.

The stack STK is initialized to empty (containing no items) by setting STK.TOP to 0, and is tested for empty by comparing STK.TOP to 0. The stack is full if STK.TOP is equal to MAX.

To place ("push") a new item, say X, onto the stack, we merely append it:

```
(* Push the value of X onto the stack STK:                    *)
WITH STK DO
    BEGIN (* push *)
    TOP := TOP + 1;
    VAL[TOP] := X
    END (* push *)
```

Ex. 39: Rewrite the push algorithm so that it will print an error message instead of trying to push a value onto a stack that is already full.

To remove ("pop") the top item from STK, putting its value into X, we can use:

```
(* Pop the top of STK into X:                                 *)
WITH STK DO
    BEGIN (* pop *)
    X := VAL[TOP];
    TOP := TOP - 1
    END (* pop *)
```

†*Ex. 40: Rewrite the algorithm so that an empty stack can not be popped.*

PROGRAM FOR READING

```
 1  PROGRAM  CHOICES( INPUT, OUTPUT );
 2
 3  (* PURPOSE:                                                    *)
 4  (*    Grade multiple-choice exams and report the results.     *)
 5  (*                                                             *)
 6  (* PROGRAMMER:  David V. Moffat                                *)
 7  (*                                                             *)
 8  (* INPUT:                                                      *)
 9  (*    The number of questions on the exam, the answer          *)
10  (*    key, and the responses of each student (probably         *)
11  (*    all of which was produced by a mark sense reader.)       *)
12  (*       line 1: n (the number of questions on the exam)       *)
13  (*       line 2: KEYKEYKEYAAAAAAAA... (the answer key)         *)
14  (*       line 3: xxxyyzzzzRRRRRRRR... (student responses)      *)
15  (*       line 4 to number of students: same as line 3.         *)
16  (*    In line 2, the As represent the correct answers.         *)
17  (*    In line 3, xxxyyzzzz represents an ID number.            *)
18  (*    The Rs represent that student's exam responses.          *)
19  (*                                                             *)
20  (* OUTPUT:                                                     *)
21  (*    An echo of the input, including each student's           *)
22  (*    score on the exam.                                       *)
23  (*    A list of scores and ID numbers in descending order      *)
24  (*    by score.                                                *)
25  (*    A summary showing the number of students, average        *)
26  (*    score, and the number of times that each question        *)
27  (*    was answered incorrectly (to show their relative         *)
28  (*    difficulty).                                             *)
29  (*                                                             *)
30  (* ASSUMPTIONS AND LIMITATIONS:                                *)
31  (*    The number of questions must be given correctly.         *)
32  (*    The maxium number of questions is 50.                    *)
33  (*    Correct answers can be in the range 1..9.                *)
34  (*    A response of "0" means "no response."                   *)
35  (*    The maximum number of students is 50.                    *)
36  (*                                                             *)
37  (* ERROR CHECKS AND RESPONSES:                                 *)
38  (*    None.                                                    *)
39
```

```
40  (* ALGORITHM:                                                *)
41  (*    Initialize counters.                                   *)
42  (*    Read and echo the # of questions and answer key.       *)
43  (*    Print titles for the input echo.                       *)
44  (*    Read and score student exam responses until EOF:       *)
45  (*       Count the student.                                  *)
46  (*       Get, echo, and save the ID.                         *)
47  (*       Get and echo responses.                             *)
48  (*       Find the student's score.                           *)
49  (*       Store the score for this ID.                        *)
50  (*    Sort the scores and IDs.                               *)
51  (*    Print titles and the ordered scores list.              *)
52  (*    Print a summary.                                       *)
53  (*    Show how often each question was missed.               *)
54
55  CONST
56     MAXQUESTIONS = 50;         (* Maximum # of questions.     *)
57     MAXSTUDENTS = 50;          (* Maximum # of students.      *)
58
59  TYPE
60     QINDEX = 1..MAXQUESTIONS;(* Question index & counter.   *)
61     QCOUNT = 0..MAXQUESTIONS;
62     SINDEX = 1..MAXSTUDENTS; (* Student index and counter. *)
63     SCOUNT = 0..MAXSTUDENTS;
64     SOCSECNUM = PACKED ARRAY[1..9] OF CHAR;
65                                (* String for ID numbers.      *)
66
67  VAR
68     ANSWER : ARRAY[QINDEX] OF CHAR;
69                                (* The answer key.             *)
70     RESPONSE : ARRAY[QINDEX] OF CHAR;
71                                (* One student's responses.    *)
72     TIMESWRONG : ARRAY[QINDEX] OF INTEGER;
73                                (* "Wrong answer" counters.    *)
74     QUESTION : QINDEX;         (* Index into above lists.     *)
75     NUMQUESTIONS : QCOUNT;     (* # of questions on exam.     *)
76     ID : ARRAY[SINDEX] OF SOCSECNUM;
77                                (* IDs of all examinees.       *)
78     THISID : SOCSECNUM;        (* One student's ID number.    *)
79     SCORE : ARRAY[SINDEX] OF INTEGER;
80                                (* Scores of all examinees.    *)
81     THISSCORE : INTEGER;       (* One student's exam score.   *)
82     STUDENT : SINDEX;          (* Index into ID and SCORE.    *)
83     NUMSTUDENTS : SCOUNT;      (* Number of examinees.        *)
84     CH : CHAR;                 (* One character of input.     *)
85     I : INTEGER;               (* For loop index.             *)
86     SUM : INTEGER;             (* Sum of all scores.          *)
87     POSN : INTEGER;            (* Loop index used in sort.    *)
88
```

```
 89   BEGIN (* CHOICES *)
 90
 91   PAGE( OUTPUT );
 92   WRITELN( 'Test Scoring Program' );
 93   WRITELN;
 94
 95   (* Initialize counters for wrong responses:                  *)
 96   FOR QUESTION:=1 TO MAXQUESTIONS DO
 97      TIMESWRONG[QUESTION] := 0;
 98
 99   (* Get and echo the number of questions:                     *)
100   READLN( NUMQUESTIONS );
101   WRITELN( 'The exam has ', NUMQUESTIONS:1, ' questions.' );
102   WRITELN;
103
104   (* Get and echo the answer key:                              *)
105   WRITELN( 'The answer key is: ' );
106   FOR I:=1 TO 9 DO
107      BEGIN (* skip ID *)
108      READ( CH );
109      WRITE( CH )
110      END; (* skip ID *)
111   WRITE( '   ' );
112   FOR QUESTION:=1 TO NUMQUESTIONS DO
113      BEGIN (* each answer *)
114      READ( CH );
115      WRITE( CH );
116      ANSWER[QUESTION] := CH
117      END; (* each answer *)
118   READLN;
119   WRITELN;
120   WRITELN;
121
122   (* Print titles for the input data echo:                     *)
123   WRITELN( 'Student     ', 'Responses', ' ':(NUMQUESTIONS-9),
124            'Score':7 );
125
```

```
126  (* Read and score student exam responses until EOF:         *)
127  NUMSTUDENTS := 0;
128  WHILE NOT EOF(INPUT) DO
129     BEGIN (* each student *)
130
131     (* Count this student:                                       *)
132     NUMSTUDENTS := NUMSTUDENTS + 1;
133
134     (* Get, echo, and save the ID:                               *)
135     FOR I:=1 TO 9 DO
136        READ( THISID[I] );
137     WRITE( THISID, '   ' );
138     ID[NUMSTUDENTS] := THISID;
139
140     (* Get and echo this student's responses:                    *)
141     FOR QUESTION:=1 TO NUMQUESTIONS DO
142        BEGIN (* each response *)
143        READ( CH );
144        WRITE( CH );
145        RESPONSE[QUESTION] := CH
146        END; (* each response *)
147     READLN;
148
149     (* Score this exam (compare the responses with the          *)
150     (* answer key), keeping running totals of the number        *)
151     (* of times each question was answered incorrectly:         *)
152     THISSCORE := 0;
153     FOR QUESTION:=1 TO NUMQUESTIONS DO
154        IF RESPONSE[QUESTION] = ANSWER[QUESTION] THEN
155           THISSCORE := THISSCORE + 1
156        ELSE
157           TIMESWRONG[QUESTION] := TIMESWRONG[QUESTION] + 1;
158     WRITELN( THISSCORE:7 );
159
160     (* Store the score in parallel with the ID:                 *)
161     SCORE[NUMSTUDENTS] := THISSCORE
162     END; (* each student *)
163
```

```
164  (* Sort the SCOREs (and IDs) into descending order by          *)
165  (* SCOREs, using a selection sort:                              *)
166  FOR STUDENT:=1 TO NUMSTUDENTS-1 DO
167     BEGIN (* each sublist *)
168     THISSCORE := SCORE[STUDENT];
169     THISID := ID[STUDENT];
170     POSN := STUDENT;
171     (* Find the largest score in the sublist:                   *)
172     FOR I:=STUDENT+1 TO NUMSTUDENTS DO
173        IF SCORE[I] > THISSCORE THEN
174           BEGIN (* save info *)
175           THISSCORE := SCORE[I];
176           THISID := ID[I];
177           POSN := I
178           END; (* save info *)
179     (* Swap the largest to the front of the sublist:           *)
180     SCORE[POSN] := SCORE[STUDENT];
181     SCORE[STUDENT] := THISSCORE;
182     ID[POSN] := ID[STUDENT];
183     ID[STUDENT] := THISID
184     END; (* each sublist *)
185
186  (* Print titles and the ordered SCOREs (and IDs):              *)
187  WRITELN;
188  WRITELN( 'List of scores in descending order:' );
189  WRITELN;
190  WRITELN( 'ID number':16, 'Score':7 );
191  FOR STUDENT:=1 TO NUMSTUDENTS DO
192     WRITELN( ID[STUDENT]:16, SCORE[STUDENT]:7 );
193  WRITELN;
194
195  (* Print a summary:                                             *)
196  WRITELN( NUMSTUDENTS:1, ' students took the exam.' );
197
198  (* Find and print the average score:                            *)
199  SUM := 0;
200  FOR STUDENT:=1 TO NUMSTUDENTS DO
201     SUM := SUM + SCORE[STUDENT];
202  WRITELN( 'The average score was ', (SUM/NUMSTUDENTS):4:1 );
203  WRITELN;
204
205  (* Show how often each question was missed:                     *)
206  WRITELN( 'Incorrect responses per question:' );
207  WRITELN;
208  FOR QUESTION:=1 TO NUMQUESTIONS DO
209     WRITELN( QUESTION:9, TIMESWRONG[QUESTION]:4 );
210  WRITELN;
211  WRITELN( 'End of program.' );
212  PAGE( OUTPUT )
213  END. (* CHOICES *)
```

```
Test Scoring Program

The exam has 15 questions.

The answer key is:
KEYKEYKEY    314152453532342

Student      Responses          Score
314159265    324152443552442      11
358979327    514252451132342      11
946264338    314155453532342      14
327950243    314152453532342      15
653543383    314152453212352      12
589794626    312153433532342      12
433832795    314152453532342      15
243141592    354152453532342      14
933650166    214152153233442      10
626327907    314132253532342      13

List of scores in descending order:

       ID number  Score
       327950243     15
       433832795     15
       946264338     14
       243141592     14
       626327907     13
       589794626     12
       653543383     12
       314159265     11
       358979327     11
       933650166     10

10 students took the exam.
The average score was 12.7

Incorrect responses per question:

         1   2
         2   2
         3   1
         4   1
         5   1
         6   2
         7   2
         8   2
         9   1
        10   3
        11   2
        12   1
        13   2
        14   1
        15   0

End of program.
```

PROGRAM EXERCISES

1. Show the changes necessary to print an asterisk under each incorrect response, like this:

```
653543383    314152453212352
                    **   *
626327907    314132253532342
                 *  *
```

2. Show how to check the "ID" of the supposed answer key to be sure that it is "KEYKEYKEY." Print an error message and stop the program if this ID is incorrect.

3. A grading policy that is sometimes believed to discourage guessing on multiple choice exams is to *subtract* 1/(n-1), where n is the number of choices, from the total score for each incorrect response, add 1 for correct responses, and do nothing for unanswered (0 response) questions. Show all the changes necessary to implement this policy.

4. The WRITELN at line 123 will cause an error if NUMQUESTIONS is less than or equal to 9. Show all the changes necessary to prevent an error.

5. Show all the changes necessary to find and print each score (and the average) as a percentage as well as a raw score.

6. The assumptions and limitations section of the program header documentation says correct answers can be in the range 1..9. Show all the changes necessary to the program (and documentation) so that it will check the responses for validity, report any problems, and assume "no response" for any invalid response.

Part IV: Matrix Algorithms

INTRODUCTION

Just as one-dimensional arrays are usually thought of as lists, matrices—two-dimensional arrays—are most commonly used to represent *tables* of information. Matrices are also mathematical objects, but, since they are seldom used that way in general programming, mathematical matrix algorithms will not be included here. Arrays of three or more dimensions are also possible, but they are hardly ever used.

Most of the declarations we will need are presented here:

```
CONST
   MAXROWS = ... ;          (* Maximum row index (integer).   *)
   MAXCOLS = ... ;          (* Maximum column index (integer).*)

TYPE
   ITEM = ... ;             (* Assume integer for now.        *)
   ONEROW = ARRAY[1..MAXCOLS] OF ITEM;
                            (* A row is a simple list.        *)
   TABLE = ARRAY[1..MAXROWS] OF ONEROW;
                            (* A table is a list of lists.    *)
   ONECOLUMN = ARRAY[1..MAXROWS] OF ITEM;
                            (* A column may be a simple list. *)

VAR
   M : TABLE;               (* A table of items.              *)
   NROWS : 0..MAXROWS;      (* Current number of rows in M.   *)
   NCOLS : 0..MAXCOLS;      (* Current number of columns in M.*)
   X : ITEM;                (* Any possible table entry for M.*)
   ROW : 1..MAXROWS;        (* Row index for M.               *)
   COL : 1..MAXCOLS;        (* Column index for M.            *)
   COLSUM : ONEROW;         (* The sums of M's columns.       *)
   ROWSUM : ONECOLUMN;      (* The sums of M's rows.          *)
   SUM : ITEM;              (* The overall sum of items in M. *)
```

The table M and the variables for the sums can be diagrammed so that the relationships of the two index variables to all three arrays are shown:

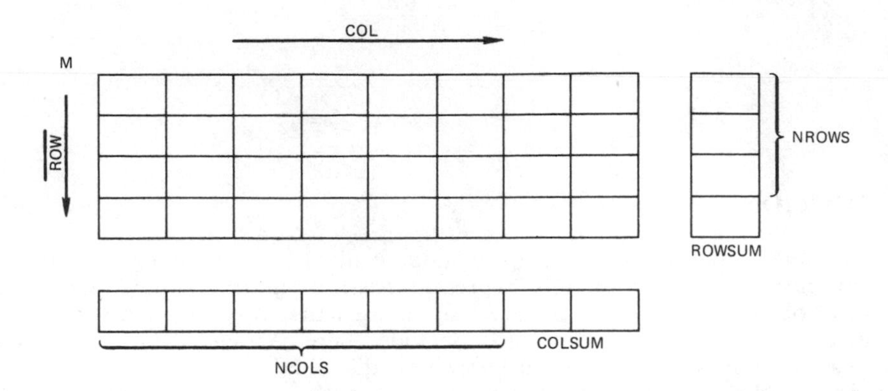

Separate index variables for ROWSUM and COLSUM may not be necessary, since ROW and COL are of the proper type. M has been carefully declared so that it can *also* be thought of as a list of lists (a list whose elements are whole lists):

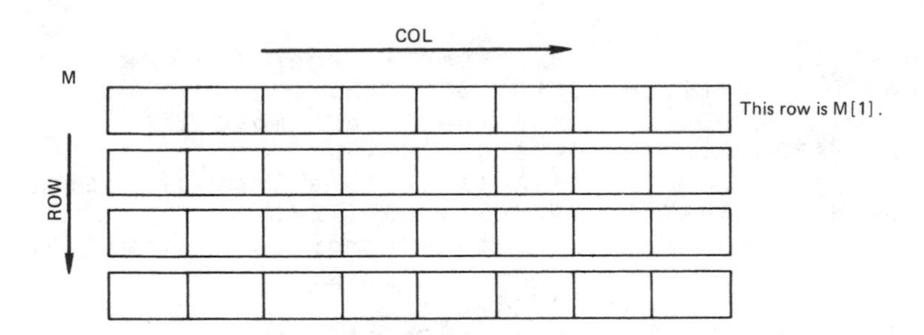

INPUT AND OUTPUT

There are many ways to input and output table values. Most of these algorithms use input or produce output that looks like one of the diagrams of the table just shown.

Assume, therefore, that each line of input contains only the values for one row of the table M.

If an extra first line of data contained two values representing the actual number of rows and columns to be read, then:

```
(* Read the size of M, then NROWS lines of NCOLS items:   *)
READ( NROWS, NCOLS );
FOR ROW:=1 TO NROWS DO
   BEGIN (* each row *)
   FOR COL:=1 TO NCOLS DO
      READ( M[ROW,COL] );
   READLN
   END (* each row *)
```

†*Ex. 1: Why is it useful (but not necessary) to have the READLN as shown in this algorithm? (Hint: How many values does each line actually contain?)*

If the extra first line of data contained only the width of the table and there were an indeterminate number of rows, the table could be input this way:

```
(* Read M's width, then its rows of NCOLS items until EOF:*)
READ( NCOLS );
NROWS := 0;
WHILE NOT EOF(INPUT) DO
   BEGIN (* each row *)
   NROWS := NROWS + 1;
   FOR COL:=1 TO NCOLS DO
      READ( M[NROWS,COL] );
   READLN
   END (* each row *)
```

Ex. 2: Write an algorithm to input values that are arranged this way: the first row is terminated by a sentinel; all rows are the same width; the last row is followed by EOF.

Simple output is quite straightforward:

```
(* Print M as a table:                                        *)
FOR ROW:=1 TO NROWS DO
   BEGIN (* each row *)
   FOR COL:=1 TO NCOLS DO
      WRITE( M[ROW,COL] );
   WRITELN
   END (* each row *)
```

More typically, the table would be labeled with titles, column numbers, and row numbers:

```
(* Print the table M with appropriate labels:                 *)

(* Print a simple title over the table:                       *)
WRITELN( '-TABLE M-':10 );
WRITELN;

(* Print the column headings (column index numbers):          *)
WRITE( ' ':4 );
FOR COL:=1 TO NCOLS DO
   WRITE( COL:6 );
WRITELN;

(* Print the table, with row numbers:                         *)
FOR ROW:=1 TO NROWS DO
   BEGIN (* each row *)
   WRITE( ROW:3, ':' );
   FOR COL:=1 TO NCOLS DO
      WRITE( M[ROW,COL]:6 );
   WRITELN
   END (* each row *)
```

Ex. 3: Show how you would print the table M so that each row of values is outlined like this:

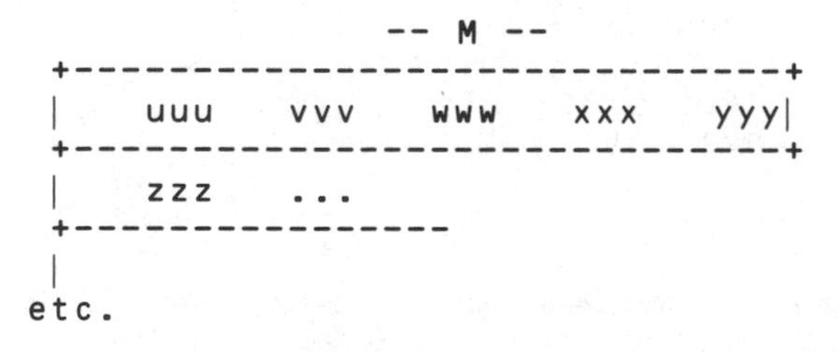

```
               -- M --
+--------------------------------+
|    uuu    vvv    www    xxx   yyy|
+--------------------------------+
|    zzz    ...
+----------------
|
etc.
```

Tables or matrices often contain more values in each row than will fit on one output line, and so the contents of each row must be divided among several lines. Given this variable:

```
COUNT : INTEGER;      (* # of values on one row so far. *)
```

This algorithm prints each row on as many lines as necessary, indenting all lines after the first for each row:

```
(* Print the table M, splitting long rows across lines:   *)
FOR ROW:= 1 TO NROWS DO
   BEGIN (* each row *)

   (* Print the row index:                                 *)
   WRITELN( ROW:3, ':');

   (* Print the contents of the row:                       *)
   COUNT := 0;
   FOR COL:=1 TO NCOLS DO
      BEGIN (* each value *)
      COUNT := COUNT + 1;
      (* Print only 10 values per line:                    *)
      IF COUNT > 10 THEN
         BEGIN (* start new line *)
         WRITELN;
         WRITE( ' ':4 );
         COUNT := 1
         END; (* start new line *)
      WRITE( M[ROW,COL] )
      END; (* each value *)

   WRITELN
   END (* each row *)
```

BASIC CALCULATIONS

Many calculations can be done on the items in a table. Algorithms perform these calculations in one of three ways: by row, by column, or overall. This section shows the three kinds of sums that might be calculated.

We can find the overall sum of the items in a table this way:

```
(* Find the overall sum of the items in the table M:          *)
SUM := 0;
FOR ROW:=1 TO NROWS DO
   FOR COL:=1 TO NCOLS DO
      SUM := SUM + M[ROW,COL]
```

The individual sums of the rows can be found like this:

```
(* Find and store the row sums of the table M:                *)
FOR ROW:=1 TO NROWS DO
   BEGIN (* each row *)
   ROWSUM[ROW] := 0;
   FOR COL:=1 TO NCOLS DO
      ROWSUM[ROW] := ROWSUM[ROW] + M[ROW,COL]
   END (* each row *)
```

†*Ex. 4: How would you find the row sums if you only needed to print them in the positions shown in the diagram?*

†*Ex. 5: (a) Write an algorithm to read rows until EOF, where each row can be any width up to MAXCOLS but is terminated by a sentinel (the same sentinel value is used for all rows).* †*(b) How would you keep track of the width of each row? (There are three ways to do this.) (c) How would you now sum the rows?*

The individual column sums can be found by turning the loops "inside out":

```
(* Find and store the column sums of the table M:             *)
FOR COL:=1 TO NCOLS DO
   BEGIN (* each column *)
   COLSUM[COL] := 0;
   FOR ROW:=1 TO NROWS DO
      COLSUM[COL] := COLSUM[COL] + M[ROW,COL]
   END (* each column *)
```

†*Ex. 6: How would you find the column sums if you did not turn the FOR loops inside out?*

Ex. 7: Rewrite the row-summing algorithm, using SUM, so that only one reference to ROWSUM is made.

Ex. 8: What is the shortest algorithm by which the row, column, and overall sums can be found and all values (including the table) displayed as in the diagram?

OTHER OPERATIONS ON TABLES

So far, tables have been treated as two-dimensional arrays of simple values. In many algorithms, however, tables are manipulated as if they were lists whose elements are also lists. The types ONEROW and TABLE have purposely been defined so that this manipulation is possible.

Since these tables are also lists, many of the list algorithms presented in Part III apply directly to tables. As examples, we can:

shift rows down

insert a new row

delete a row

remove duplicate rows

merge two tables

split a table into two

These algorithms are all quite straightforward. That fact is illustrated with a simple example. Assume that we have this variable:

```
    HOLD : ONEROW;          (* Holds a whole row temporarily. *)
```

Then, for example, two entire rows can easily be swapped:

```
(* Exchange two consecutive rows of M:                        *)
HOLD := M[ROW];
M[ROW] := M[ROW+1];
M[ROW+1] := HOLD
```

†*Ex. 9: To exchange two columns of values requires a loop. Write an algorithm to make the exchange.*

At the same time, we can continue to refer to individual items in each row. In particular, often one item in the same column of each row is of primary importance. This is the "primary column" of the table. The other values in the row are related to the value in the primary column but are of secondary interest. So, assuming these declarations:

```
CONST
   PRIMARY = ...;          (* Index of the primary column.    *)

VAR
   INORDER : BOOLEAN;      (* TRUE if the table is ordered.  *)
   LIMIT : 1..MAXROWS;     (* Limit for each sort pass.       *)
```

then we might order the rows of M relative to the values contained only in the primary column:

```
(* Bubble sort: Order M relative to its primary column:    *)
LIMIT := NROWS;
REPEAT
   (* Assume that the table is already ordered:            *)
   INORDER := TRUE;

   (* Limit the comparison to the next shorter sublist:    *)
   LIMIT := LIMIT - 1;

   (* Compare column elements and order all pairs of rows:*)
   FOR ROW:=1 TO LIMIT DO
      BEGIN (* each row pair *)
      (* Put this pair in ascending order:                 *)
      IF M[ROW,PRIMARY] > M[ROW+1,PRIMARY] THEN
         BEGIN (* swap *)
         (* The table is not yet in order:                 *)
         INORDER := FALSE;
         HOLD := M[ROW];
         M[ROW] := M[ROW+1];
         M[ROW+1] := HOLD
         END (* swap *)
      END (* each row pair *)

UNTIL INORDER
```

Ex. 10: Write a sort to order M relative to the contents of two columns. That is, if a pair of items in the PRIMARY column are equal, break the "tie" by comparing items in the next column.
Ex. 11: Write a sort to order M so that each row is ordered within itself and the rows are ordered with respect to the contents of a given column.

Tables are also searched, either as lists of rows or as matrices of separate values. Assume that we have these variables:

```
FOUND : BOOLEAN;       (* TRUE if search is a success.   *)
ROWPOS : INTEGER;      (* Row index for searching table. *)
COLPOS : INTEGER;      (* Column index for the search.   *)
```

First, a table that is thought of as a list of rows may have a primary column relative to which it is sorted or to which the rest of the row values correspond. That column, PRIMARY, would be searched to find the row represented by the value of a given item X:

```
(* Linear search of table M's primary column for X's row: *)
ROWPOS := 1;
FOUND := FALSE;
WHILE (ROWPOS <= NROWS) AND NOT FOUND DO
   IF M[ROWPOS,PRIMARY] = X THEN
      FOUND := TRUE
   ELSE
      ROWPOS := ROWPOS + 1
(* If FOUND, ROWPOS is X's position in the PRIMARY column.*)
```

Ex. 12: Recall the list comparison algorithm in the List Operations section of Part III. Write an algorithm to search M for a row equal to XROW of type ONEROW.

We may also think of the table as consisting of individual items, rather than as a list of rows. In this case we may want to find both the row and the column at which the value of X is stored:

```
(* Linear search of matrix M to find the position of X:   *)
FOUND := FALSE;
ROWPOS := 1;
WHILE (ROWPOS <= NROWS) AND NOT FOUND DO
   BEGIN (* each row *)

   (* Search this row for X:                               *)
   COLPOS := 1;
   WHILE (COLPOS <= NCOLS) AND NOT FOUND DO
      BEGIN (* each column *)
      IF M[ROWPOS,COLPOS] = X THEN
         FOUND := TRUE
      ELSE
         COLPOS := COLPOS + 1
      END; (* each column *)

   (* Increment to search the next row only if necessary: *)
   IF NOT FOUND THEN
      ROWPOS := ROWPOS + 1

   END (* each row *)
(* If FOUND, then X is at M[ROWPOS,COLPOS].                 *)
```

†*Ex. 13: Given a matrix of integers in which the values within each row are in ascending order and increasing from row to row, write a faster search for the value of X. (Hint: Recall the quadratic search of Part III.)*

OTHER KINDS AND USES OF TABLES

Homogeneous matrices, or lists of lists, are not the only kinds of tables—perhaps not even the most important kinds of tables. A list of any other structured type—records, sets, or strings—can also be thought of as a table.

A list of strings, however, is usually manipulated simply as a list, not as a table of characters. Lists are covered in Part III, strings in Part V.

Lists of sets are not common. Certainly such a table would not be "ordered" in any way, except that each set would be known by where it appeared in the list.

Tables consisting of lists of records *are* very common and very important.

Assume that we have these declarations instead of the ones given in the Introduction:

```
CONST
   MAXROWS = ...;          (* Maximum # of rows in a table.  *)

TYPE
   KEYITEM = ...;          (* Assume integer, for example.   *)
   ONEROW =                (* A row of a table (that is, an  *)
      RECORD               (* element of a list).            *)
      KEY : KEYITEM;       (* See discussion.                *)
      INFO : ...           (* See discussion.                *)
      END;
                           (* A table is a list of records:  *)
   TABLE = ARRAY[1..MAXROWS] OF ONEROW;

VAR
   T : TABLE;
   ROW : 1..MAXROWS;       (* Index into the table T.        *)
   NROWS : 0..MAXROWS;     (* Actual number of records used. *)
   X : KEYITEM;            (* Any possible key value.        *)
   POSN : INTEGER;         (* Index for searches and sorts.  *)
   I : 1..MAXROWS;         (* Index for searches and sorts.  *)
   HOLD : ONEROW;          (* For swapping rows.             *)
   FOUND : BOOLEAN;        (* TRUE if search is successful.  *)
```

Each record in the table T contains information of some kind (in the field called INFO here) and a special field, typically called KEY, that serves as a "key" to the information. For example, in a table of drivers' licenses, the key might be the license number; the name, address, and so on, corresponding to a given number, are found in the record containing that number.

The input and output of these tables is quite straightforward:

```
(* Input and echo rows of table T until EOF:                *)
NROWS := 0;
WHILE NOT EOF(INPUT) DO
   BEGIN (* each row *)
   NROWS := NROWS + 1;
   WITH T[NROWS] DO
      BEGIN (* get fields *)
      READ( KEY );
      READLN( INFO );
      WRITELN( KEY, INFO )
      END (* get fields *)
   END (* each row *)
```

The table is most often ordered with respect to the key fields, just as the rows of a matrix might be ordered with respect to the primary column. Thus, if X contains the value of some key, we can search for this key in the table and find the corresponding information in the INFO field associated with it:

```
(* Linear search of table T to find the record that         *)
(* contains the key value X:                                 *)
POSN := 1;
FOUND := FALSE;
WHILE (POSN <= NROWS) AND NOT FOUND DO
   BEGIN (* each record *)
   IF T[POSN].KEY < X THEN
      POSN := POSN + 1
   ELSE
      FOUND := TRUE
   END (* each record *)
(* Now POSN points to the record whose KEY should equal X.*)
```

Ex. 14: Rewrite this algorithm as a binary or quadratic search.

The table may be put into order any number of ways, either during input or by using one of the many sorting algorithms.

Ex. 15: Write an algorithm to build an ordered table T from unordered input, assuming a simple type for INFO.

†*Ex. 16: Show how to sort an unordered table T with one of the common sorts. Use HOLD for swapping rows (entire records).*

If the individual records and the table as a whole are very large, the common sorts may execute very slowly because of all the data that is moved around by the swaps or inserts. The so-called "tag sort" was developed to overcome this problem. It uses an array of "tags" (table indices or pointers) that point to the rows of the table. All references to the table's rows are made indirectly, through the tags. The *tags* can be sorted so that they point to the rows in increasing order by keys.

Given this tag array and swapping variable:

```
                              (* Tag array, of table indices.    *)
   TAG : ARRAY[1..MAXROWS] OF 1..MAXROWS;
   TEMP : 1..MAXROWS;         (* For swapping TAG values.        *)
```

we can sort and (for example) print the table T this way (you may wish first to look at the diagram of tag and table values that follows Ex. 17):

```
(* Tag sort (using a selection sort) of table T:                *)

(* The tags must contain sequential values at first:            *)
FOR ROW:=1 TO NROWS DO
   TAG[ROW] := ROW;

(* The actual sorting.  The keys are compared by way of         *)
(* the tags, but the tags are swapped to show the order.        *)
FOR ROW:=1 TO NROWS-1 DO
   BEGIN (* each sublist *)

   (* Find the smallest key in the sublist of T:                *)
   X := T[TAG[ROW]].KEY;
   POSN := ROW;
   FOR I:=ROW+1 TO NROWS DO
      IF T[TAG[I]].KEY < X THEN
         BEGIN (* save smallest *)
         X := T[TAG[I]].KEY;
         POSN := I
         END; (* save smallest *)

   (* Put smallest's tag at the beginning of the tags for *)
   (* this sublist:                                            *)
   TEMP := TAG[ROW];
   TAG[ROW] := TAG[POSN];
   TAG[POSN] := TEMP
   END; (* each sublist *)

(* Print the keys in sorted order:                              *)
FOR ROW:=1 TO NROWS DO
   WRITELN( T[TAG[ROW]].KEY )
```

Ex. 17: Write a tag sort based upon the insertion sort.

This diagram illustrates the tag values before and after a tag sort:

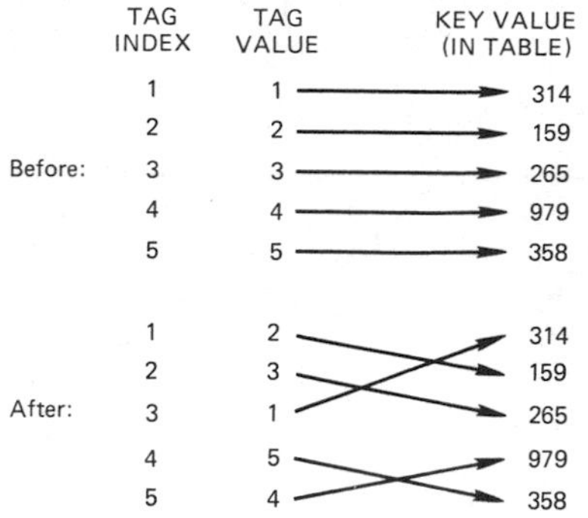

	TAG INDEX	TAG VALUE	KEY VALUE (IN TABLE)
	1	1 →	314
	2	2 →	159
Before:	3	3 →	265
	4	4 →	979
	5	5 →	358
	1	2	314
	2	3	159
After:	3	1	265
	4	5	979
	5	4	358

Note that after the sort T[TAG[1]].KEY is less than T[TAG[2]].KEY, and so on.

Matrices and tables are also used as tabulated functions. A table can be thought of as a function of one variable (the row index) that has an entire row (a list, record, set, or string) as its value. This is the only way to create a function that returns structured values. On the other hand, a matrix can be used as a function of two variables (row and column indices) that returns simple values. This application is particularly useful because of the many possible combinations of the basic index types (16, in fact).

This simple example shows a typical use of a table as a function:

```
TYPE
   RESPONSE = ( NO, YES );
   WORD = PACKED ARRAY[1..3] OF CHAR;

VAR
   ANSWER : RESPONSE;
   STRINGOF : ARRAY[RESPONSE] OF WORD;
                      (* "Returns" a string, given any  *)
                      (* value of type RESPONSE.        *)

 ...

(* Initialize the function:                             *)
STRINGOF[NO ] := 'No ';
STRINGOF[YES] := 'Yes';
 ...

(* Use the function:                                    *)
ANSWER := ...;
WRITE( STRINGOF[ANSWER] )
```

PROGRAM FOR READING

```
 1  PROGRAM  CHECKING( INPUT, OUTPUT );
 2
 3  (* PURPOSE:                                                  *)
 4  (*    Generate a statement for a checking account.           *)
 5  (*                                                           *)
 6  (* PROGRAMMER:  David V. Moffat                              *)
 7  (*                                                           *)
 8  (* INPUT:                                                    *)
 9  (*    Transactions applied to the account, one per line,     *)
10  (*    formatted this way:                                    *)
11  (*    col.  1- 3: a three-digit check number, or             *)
12  (*                '000' for a noncheck withdrawal, or        *)
13  (*                'DEP' to identify a deposit, or            *)
14  (*                'BAL' to identify the initial balance.     *)
15  (*    col.     4: blank                                      *)
16  (*    col.  5-11: the amount of the transaction (dddd.cc)    *)
17  (*    col.    12: blank                                      *)
18  (*    col. 13-18: date of transaction (mmddyy)               *)
19  (*    col.    19: blank                                      *)
20  (*    col. 20-29: category or explanation (optional)         *)
21  (*    The BAL transaction, if any, should be first.          *)
22  (*                                                           *)
23  (* OUTPUT:                                                   *)
24  (*    An echo of the input, with a running balance.          *)
25  (*    Summaries for the credits and the debits, showing      *)
26  (*    the number of transactions in each category, the       *)
27  (*    amount of money and percent of total represented by    *)
28  (*    each category, and the totals.                         *)
29  (*                                                           *)
30  (* ASSUMPTIONS AND LIMITATIONS:                              *)
31  (*    The initial balance is $0.00 if not specified.         *)
32  (*    There can be up to 25 debits and 25 credits.           *)
33  (*                                                           *)
34  (* ERROR CHECKS AND RESPONSES:                               *)
35  (*    If there are too many credits or debits, then the      *)
36  (*    transaction is ignored and a message is printed.       *)
37  (*                                                           *)
38  (* ALGORITHM:                                                *)
39  (*    Initialize the tables and the balance.                 *)
40  (*    Process the transactions, accumulating the             *)
41  (*       information into tables.                            *)
42  (*    Print summaries of the credits and debits.             *)
43
```

```
44   CONST
45      MAXSIZE = 25;              (* Maximum table size.          *)
46      NAMELENGTH = 10;           (* Size of category names.      *)
47
48   TYPE
49      CATNAME = PACKED ARRAY[1..NAMELENGTH] OF CHAR;
50                                 (* String for category names.   *)
51                                 (* Other strings:               *)
52      STRING3 = PACKED ARRAY[1..3] OF CHAR;
53      STRING6 = PACKED ARRAY[1..6] OF CHAR;
54      STRING7 = PACKED ARRAY[1..7] OF CHAR;
55
56      TRANSTYPE =                (* Transactions from input:     *)
57         RECORD
58         KIND : STRING3;         (* 'BAL', 'DEP', or check #.    *)
59         AMOUNT : REAL;          (* Amount of credit or debit.   *)
60         DATE : STRING6;         (* Date of transaction.         *)
61         WHATFOR : CATNAME       (* Source of deposit, or the    *)
62         END;                    (* reason for the expenditure.  *)
63
64      TABLEITEM =                (* Credit and debit table       *)
65         RECORD                  (* entries (for summaries):     *)
66         CATEGORY : CATNAME;     (* Credit or debit category.    *)
67         NUMBER : 0..MAXINT;     (* Number of transactions in    *)
68                                 (* this category.               *)
69         SUBTOTAL : REAL         (* Amount of money represented  *)
70         END;                    (* by this category.            *)
71
72         TABLE = ARRAY[1..MAXSIZE] OF TABLEITEM;
73                                 (* Credit and debit tables.     *)
74         COUNT = 0..MAXSIZE;     (* Current table size.          *)
75
76   VAR
77      CREDITS : TABLE;           (* Summary of incomes.          *)
78      NUMCREDITS : COUNT;        (* Current size of CREDITS.     *)
79      DEBITS : TABLE;            (* Summary of expenditures.     *)
80      NUMDEBITS : COUNT;         (* Current size of DEBITS.      *)
81      STARTINGBALANCE : REAL;    (* Defaults to $0.00.           *)
82      BALANCE : REAL;            (* Current (running) balance.   *)
83
```

```
 84   PROCEDURE  GETONE( VAR TRANSACTION : TRANSTYPE );
 85      (* Read the data for one checking account TRANSACTION  *)
 86      (* from one line of input.                             *)
 87
 88      VAR
 89         CH : CHAR;           (* One character of input.     *)
 90         I : INTEGER;         (* Loop index.                 *)
 91
 92      BEGIN (* GETONE *)
 93
 94      WITH TRANSACTION DO
 95         BEGIN (* each TRANSACTION *)
 96         (* Get the kind of the transaction:                 *)
 97         FOR I:=1 TO 3 DO
 98           BEGIN
 99           READ( CH );
100           KIND[I] := CH
101           END;
102         (* Get the dollar amount, and skip the blank:       *)
103         READ( AMOUNT, CH );
104         (* Get the date of the transaction, and skip the    *)
105         (* blank (if any):                                  *)
106         FOR I:=1 TO 6 DO
107            BEGIN
108            READ( CH );
109            DATE[I] := CH
110            END;
111         IF NOT EOLN(INPUT) THEN
112            READ( CH );
113         (* Get the category, if any, and finish the line:   *)
114         FOR I:=1 TO NAMELENGTH DO
115            IF NOT EOLN(INPUT) THEN
116               BEGIN
117               READ( CH );
118               WHATFOR[I] := CH
119               END
120            ELSE
121               WHATFOR[I] := ' ';
122         READLN
123         END (* each TRANSACTION *)
124
125      END; (* GETONE *)
126
127
128   PROCEDURE PUTONE( TRANSACTION : TRANSTYPE; BALANCE : REAL );
129      (* Print a TRANSACTION and the current BALANCE.        *)
130
131      BEGIN (* PUTONE *)
132
133      WITH TRANSACTION DO
134         WRITE( KIND:4, AMOUNT:9:2, DATE:7, WHATFOR:11 );
135      WRITELN( BALANCE:9:2 )
136      END; (* PUTONE *)
```

```
137
138
139  PROCEDURE  INSERT( TRANSACTION : TRANSTYPE;
140                     VAR ANYTABLE : TABLE;
141                     VAR SIZE : COUNT;  POSN : INTEGER );
142     (* Insert a new TRANSACTION into ANYTABLE (which    *)
143     (* contains SIZE entries) at position POSN.         *)
144
145     VAR
146        I : INTEGER;      (* Loop index.                 *)
147
148     BEGIN (* INSERT *)
149
150     IF POSN > MAXSIZE THEN
151        WRITELN( '***Error: too many different ',
152                 'categories--this one ignored.' )
153     ELSE
154        BEGIN (* insert or append *)
155        FOR I:=SIZE DOWNTO POSN DO
156           ANYTABLE[I+1] := ANYTABLE[I];
157        SIZE := SIZE + 1;
158        WITH ANYTABLE[POSN], TRANSACTION DO
159           BEGIN (* initialize *)
160           CATEGORY := WHATFOR;
161           SUBTOTAL := AMOUNT;
162           NUMBER := 1
163           END (* initialize *)
164        END (* insert or append *)
165
166     END; (* INSERT *)
167
```

```
168  PROCEDURE  UPDATE( VAR ANYTABLE : TABLE;  VAR SIZE : COUNT;
169                         TRANSACTION : TRANSTYPE );
170     (* Find the category of the given TRANSACTION in          *)
171     (* ANYTABLE (containing SIZE entries) and update the      *)
172     (* corresponding entry fields.                            *)
173     (* If the category is not already there, insert it.      *)
174
175     VAR
176        POSN : INTEGER;        (* Index for the search.         *)
177        FOUND : BOOLEAN;       (* TRUE if search successful.    *)
178
179     BEGIN (* UPDATE *)
180
181     (* Search the table for the given category:               *)
182     POSN := 1;
183     FOUND := FALSE;
184     WHILE (POSN <= SIZE) AND NOT FOUND DO
185        IF TRANSACTION.WHATFOR <= ANYTABLE[POSN].CATEGORY THEN
186           FOUND := TRUE
187        ELSE
188           POSN := POSN + 1;
189     (* Update the entry (if found) or insert a new one:       *)
190     IF FOUND THEN
191        IF TRANSACTION.WHATFOR = ANYTABLE[POSN].CATEGORY THEN
192           (* The category is already at POSN in the table: *)
193           WITH ANYTABLE[POSN] DO
194              BEGIN (* increment entry *)
195              NUMBER := NUMBER + 1;
196              SUBTOTAL := SUBTOTAL + TRANSACTION.AMOUNT
197              END (* increment entry *)
198        ELSE
199           (* The category should be inserted in the middle:*)
200           INSERT( TRANSACTION, ANYTABLE, SIZE, POSN )
201     ELSE
202        (* The category must be appended:                      *)
203        INSERT( TRANSACTION, ANYTABLE, SIZE, POSN )
204
205     END; (* UPDATE *)
206
```

```
207   PROCEDURE  ACCUMULATE( VAR CREDITS : TABLE;
208                          VAR NUMCREDITS : COUNT;
209                          VAR DEBITS : TABLE;
210                          VAR NUMDEBITS : COUNT;
211                          VAR STARTINGBALANCE : REAL;
212                          VAR BALANCE : REAL );
213      (* Given the STARTINGBALANCE on the checking account,   *)
214      (* get, process, and echo the account transactions,     *)
215      (* updating the CREDITS and DEBITS tables, and the      *)
216      (* BALANCE.                                             *)
217
218      VAR
219         TRANSACTION : TRANSTYPE;
220                           (* One account transaction.    *)
221
222      BEGIN (* ACCUMULATE *)
223
224      (* Print titles and column headings for the echo:      *)
225      PAGE(OUTPUT);
226      WRITELN( 'Record of all account transactions:' );
227      WRITELN;
228      WRITELN( 'Kind', 'Amount':9, 'mmddyy':7, 'Category':11,
229               'Balance':9 );
230
231      (* Get and process transactions until end of file:     *)
232      WHILE NOT EOF(INPUT) DO
233         BEGIN (* each TRANSACTION *)
234         (* Get the data:                                    *)
235         GETONE( TRANSACTION );
236         (* Update the balance and a table, depending upon   *)
237         (* the kind of transaction it is:                   *)
238         IF TRANSACTION.KIND = 'BAL' THEN
239            BEGIN (* set balances *)
240            STARTINGBALANCE := TRANSACTION.AMOUNT;
241            BALANCE := STARTINGBALANCE
242            END (* set balances *)
243         ELSE IF TRANSACTION.KIND = 'DEP' THEN
244            BEGIN (* credit the account *)
245            UPDATE( CREDITS, NUMCREDITS, TRANSACTION );
246            BALANCE := BALANCE + TRANSACTION.AMOUNT
247            END (* credit the account *)
248         ELSE
249            BEGIN (* debit the account *)
250            UPDATE( DEBITS, NUMDEBITS, TRANSACTION );
251            BALANCE := BALANCE - TRANSACTION.AMOUNT
252            END; (* debit the account *)
253         (* Echo the transaction and the current balance:    *)
254         PUTONE( TRANSACTION, BALANCE )
255         END (* each TRANSACTION *)
256
257      END; (* ACCUMULATE *)
```

```
258
259  PROCEDURE  SORT( VAR ANYTABLE : TABLE; SIZE : COUNT );
260     (* Sort the given table ANYTABLE (containing SIZE          *)
261     (* entries) into ascending order (by the SUBTOTAL for      *)
262     (* each CATEGORY), using and insertion sort.               *)
263
264     VAR
265        ITEM : TABLEITEM;      (* For swapping table entries. *)
266        I : INTEGER;           (* Loop index.                 *)
267        POSN : INTEGER;        (* Index for searching.        *)
268        FOUND : BOOLEAN;       (* TRUE if search successful.  *)
269
270     BEGIN (* SORT *)
271
272     FOR I:=2 TO SIZE DO
273        BEGIN (* each entry *)
274        ITEM := ANYTABLE[I];
275        POSN := I-1;
276        FOUND := FALSE;
277        WHILE (POSN >= 1) AND NOT FOUND DO
278           BEGIN (* search and shift *)
279           IF ITEM.SUBTOTAL < ANYTABLE[POSN].SUBTOTAL THEN
280              BEGIN (* shift *)
281              ANYTABLE[POSN+1] := ANYTABLE[POSN];
282              POSN := POSN - 1
283              END (* shift *)
284           ELSE
285              FOUND := TRUE
286           END; (* search and shift *)
287        ANYTABLE[POSN+1] := ITEM
288        END (* each entry *)
289
290     END; (* SORT *)
291
```

```
292  PROCEDURE  PRINTTABLE( ANYTABLE : TABLE; SIZE : COUNT );
293     (* Print the given table, ANYTABLE, containing SIZE      *)
294     (* entries, labeled with column headings and totals.    *)
295
296     VAR
297        I : INTEGER;          (* Loop index.                  *)
298        TOTALTRANS : INTEGER;(* Total number of trans.       *)
299        TOTALAMOUNT : REAL;  (* Total amount of trans.       *)
300
301     BEGIN (* PRINTTABLE *)
302
303     (* Find the total number and dollar amount of the       *)
304     (* transactions in the given table:                     *)
305     TOTALTRANS := 0;
306     TOTALAMOUNT := 0.0;
307     FOR I:=1 TO SIZE DO
308        WITH ANYTABLE[I] DO
309           BEGIN
310           TOTALTRANS := TOTALTRANS + NUMBER;
311           TOTALAMOUNT := TOTALAMOUNT + SUBTOTAL
312           END;
313     (* Print column headings and the table:                 *)
314     WRITELN;
315     WRITELN( ' ':10, '# of':7, ' ':10, '% of':7 );
316     WRITELN( 'Category':10, 'Items':7, 'Subtotal':10,
317              'Total':7 );
318     FOR I:=1 TO SIZE DO
319        WITH ANYTABLE[I] DO
320           WRITELN( CATEGORY:10, NUMBER:7, SUBTOTAL:10:2,
321                    (SUBTOTAL/TOTALAMOUNT*100):7:2 );
322     (* Print the totals:                                    *)
323     WRITELN;
324     WRITELN( '***Totals:':10, TOTALTRANS:7,
325              TOTALAMOUNT:10:2, (100.00):7:2 )
326
327     END; (* PRINTTABLE *)
328
329
330
331  PROCEDURE  DISPLAY( ANYTABLE : TABLE;
332                      SIZE : COUNT;
333                      TITLE : STRING7 );
334     (* Display the given table, ANYTABLE, which contains    *)
335     (* SIZE entries, in ascending order, under the TITLE.   *)
336
337     BEGIN (* DISPLAY *)
338
339     WRITELN( 'Summary of all ':20, TITLE );
340     SORT( ANYTABLE, SIZE );
341     PRINTTABLE( ANYTABLE, SIZE )
342
343     END; (* DISPLAY *)
```

```
344
345  BEGIN (* CHECKING *)
346
347  (* Initialize the tables and the balance:                        *)
348  NUMCREDITS := 0;
349  NUMDEBITS := 0;
350  STARTINGBALANCE := 0.0;
351  BALANCE := 0.0;
352
353  (* Process transactions, accumulating them into tables:   *)
354  ACCUMULATE( CREDITS, NUMCREDITS,
355              DEBITS, NUMDEBITS,
356              STARTINGBALANCE, BALANCE );
357
358  (* Print summaries of the credits and debits:                  *)
359  PAGE( OUTPUT );
360  DISPLAY( CREDITS, NUMCREDITS, 'Credits' );
361  WRITELN;
362  WRITELN;
363  WRITELN;
364  DISPLAY( DEBITS, NUMDEBITS, 'Debits ' );
365  WRITELN;
366  WRITELN;
367  WRITELN;
368  WRITELN( 'The starting balance was:', STARTINGBALANCE:9:2 );
369  WRITELN( 'The final balance is:', BALANCE:9:2 );
370  PAGE( OUTPUT )
371
372  END. (* CHECKING *)
```

```
Record of all account transactions:

Kind    Amount mmddyy   Category    Balance
 BAL    291.86 070183                291.86
 469     10.66 070283 FOODOUT        281.20
 000     25.00 070183 CASH           256.20
 DEP    241.86 070383 RESTOFPAY      498.06
 470     20.80 070383 GROCERIES      477.26
 471     26.85 070383 GROCERIES      450.41
 472    200.00 070583 RENT           250.41
 473     92.81 070583 TELEPHONE      157.60
 474     15.30 070583 ELECTRIC       142.30
 475     41.02 070583 PIANO          101.28
 476     88.33 070583 LOAN            12.95
 477    122.00 070583 AUTOINSURA    -109.05
 DEP    300.00 070583 FROMSAVES      190.95
 478      7.96 071683 FOODOUT        182.99
 DEP    246.12 071783 RESTOFPAY      429.11
 479     29.21 071783 GROCERIES      399.90
 000     25.00 070583 CASH           374.90
 000     25.00 071783 CASH           349.90
 480     18.00 071883 MOTORCLUB      331.90
 481     16.64 071883 ELECTRIC       315.26
 482     18.04 071883 FOODOUT        297.22
 DEP    400.00 072383 FROMSAVES      697.22
 DEP    275.00 072383 FROMSAVES      972.22
 483     74.54 072283 CLOTHES        897.68
 000     25.00 072383 CASH           872.68
 484      6.91 072383 HOUSE          865.77
 485     17.98 072383 GROCERIES      847.79
 486     14.77 072483 HOUSE          833.02
 487      5.56 072483 HARDWARE       827.46
 488      9.15 072483 FOODOUT        818.31
 000     25.00 072383 CASH           793.31
 489     14.32 072483 DRUGSTORE      778.99
 490      6.91 072583 FOODOUT        772.08
 491     12.97 072683 HOUSE          759.11
 492     30.00 072883 PIANO          729.11
 493      5.00 072983 GARBAGE        724.11
 DEP    270.00 073083 RESTOFPAY      994.11
 000     25.00 073183 CASH           969.11
 494      8.06 080783 FOODOUT        961.05
 495     39.40 080883 GROCERIES      921.65
 000     25.00 080583 CASH           896.65
 000      3.76 080183 SERVICE        892.89
 501      6.25 080883 GARBAGE        886.64
 502     41.02 080883 PIANO          845.62
 503     88.33 080883 LOAN           757.29
 504      6.92 081383 FOODOUT        750.37
 505     16.53 081483 GROCERIES      733.84
 000     25.00 081083 CASH           708.84
 506     66.61 081583 TELEPHONE      642.23
 DEP    240.00 081383 RESTOFPAY      882.23
```

```
      Summary of all Credits

              # of              % of
  Category   Items  Subtotal   Total
FROMSAVES        3    975.00   49.42
RESTOFPAY        4    997.98   50.58

***Totals:       7   1972.98  100.00


      Summary of all Debits

              # of              % of
  Category   Items  Subtotal   Total
SERVICE          1      3.76    0.27
HARDWARE         1      5.56    0.40
GARBAGE          2     11.25    0.81
DRUGSTORE        1     14.32    1.04
MOTORCLUB        1     18.00    1.30
ELECTRIC         2     31.94    2.31
HOUSE            3     34.65    2.51
FOODOUT          7     67.70    4.90
CLOTHES          1     74.54    5.39
PIANO            3    112.04    8.10
AUTOINSURA       1    122.00    8.82
GROCERIES        6    150.77   10.90
TELEPHONE        2    159.42   11.53
LOAN             2    176.66   12.78
CASH             8    200.00   14.47
RENT             1    200.00   14.47

***Totals:      42   1382.61  100.00


The starting balance was:   291.86
The final balance is:   882.23
```

PROGRAM EXERCISES

1. Change the names of the types STRING3, STRING6, and STRING7 to KINDTYPE, DATETYPE, and TITLETYPE throughout the program.

2. (a) What kind of search is used to find the position for a given category in a given table? (b) If a different search were used, would the insertion algorithm have to be changed?

3. Show all the changes necessary to print the '000' transactions (which, by the way, represent service charges and "automatic teller" withdrawals) as a separate summary.

4. There could be a 26th position in the tables, and UPDATE could be changed so that all transactions with new categories after the first 25 different ones will be added together into a 26th category called "OTHER," rather than discarded. Show the necessary changes.

5. Where and how should the program be changed to allow "free format" input—that is, transactions with any number of blanks between items?

6. Complete the assumptions and limitations section of the program documentation. In particular, what other limitations are implied by the data structures and the output formats?

7. Draw and label a diagram showing how the program is divided into modules. Does each module perform one task only? What are the advantages and disadvantages of requiring all modules to have parameters even if some modules are called only once?

8. Determine how the CHOICES program of Part III might be changed to store the IDs, responses, and scores in a table. What would be gained by these changes?

Part V: String Algorithms

INTRODUCTION

In Part II character-processing algorithms were discussed. Such algorithms are—and should be—used whenever character data can be processed immediately, one character at a time. Often, however, the characters must be saved in strings so that they can be processed later on. These strings are conceptually distinct from mere arrays of characters. They are manipulated by a wide range of special string-processing algorithms—the subject of this part.

The most important applications of string-processing algorithms are word- or text-processing programs. It was one such program that took the unformatted text of this book (mixed with formatting instructions or "commands") and produced neatly arranged pages.

For string algorithms we need two different kinds of strings:

```
CONST
   MAXLEN = ...;          (* Maximum # of chars in a string.*)

TYPE
   FIXEDSTRING = PACKED ARRAY[1..MAXLEN] OF CHAR;
                          (* A "fixed-length" string type.   *)

   STRING =               (* A "variable-length" string:     *)
      RECORD
      STR : FIXEDSTRING;(* The actual string.                *)
      LEN : 0..MAXLEN     (* The length being "used."        *)
      END;

VAR
   F : FIXEDSTRING;       (* Some example strings:           *)
   S : STRING;

   NULL : STRING;         (* "Empty" strings (see text):     *)
   BLANKS : FIXEDSTRING;
```

Most of the algorithms in the rest of the book are presented as procedures or functions.

Objects of type FIXEDSTRING are simple strings. When we use one, we recognize that we are using the entire string. Objects of type STRING are conceptually different: they are treated as if their lengths could vary from 0 to MAXLEN. This is done by using the LEN field to tell how much of STR is "in use" at any time:

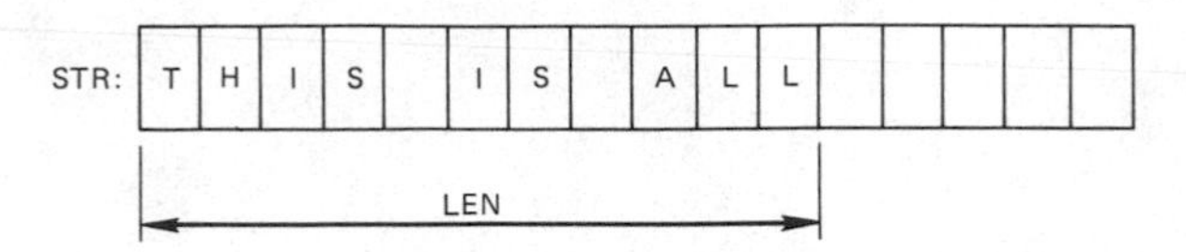

We must be careful about two things: it is our responsibility to keep LEN up to date and always to keep the "unused" portion (the part beyond LEN) of STR blank.

BLANKS and NULL are global variables used merely to initialize the other string variables. They would themselves have to be initialized at the beginning of the program:

```
PROCEDURE  INITSTRINGS( VAR BLANKS : FIXEDSTRING;
                        VAR NULL : STRING );
   (* Set up the initialization strings for later use:    *)

   VAR
      I : 1..MAXLEN;     (* FOR loop index.               *)

   BEGIN (* INITSTRINGS *)

   FOR I:=1 TO MAXLEN DO
      BLANKS[I] := ' ';

   (* Set NULL to the "empty" string (0 characters used): *)
   WITH NULL DO
      BEGIN
      STR := BLANKS;
      LEN := 0
      END

   END; (* INITSTRINGS *)
```

Now, throughout the program, other strings can be initialized when necessary:

```
F := BLANKS;
S := NULL
```

(If FIXEDSTRINGs were short enough, BLANKS could be a constant.)

In this part of the book the variable-length objects of type STRING are emphasized. Algorithms for FIXEDSTRING objects can be derived from the STRING algorithms.

INPUT AND OUTPUT

The used portion of a string typically includes only the characters significant to the problem at hand. That is, there are usually no extraneous characters such as the trailing blanks that occur at the end of fixed-length strings. Thus, algorithms for input and output must pay close attention to the LEN field.

Input is best done with an algorithm that uses the input buffer to read *only* the characters that belong in the string:

```
PROCEDURE  READSTRING( VAR S : STRING );
   (* Read characters into the string S up to, but not      *)
   (* including, the first blank:                           *)

   VAR
      CH : CHAR;          (* One character of input.        *)

   BEGIN (* READSTRING *)

   S := NULL;
   WITH S DO
      WHILE INPUT@ <> ' ' DO          (* @ is the same as ^ *)
         BEGIN (* each nonblank *)
         READ( CH );
         LEN := LEN + 1;
         STR[LEN] := CH
         END (* each nonblank *)

   END; (* READSTRING *)
```

Ex. 1: Redo READSTRING so that it accepts a given character as a parameter, then reads into S all characters up to that character or to EOLN. The given character should not be included in the string.

†*Ex. 2: Write a procedure called READQUOTED that reads a quoted string from the input, putting all but the quotes into the given string parameter. (Assume that there are no quotes within the string.)*

Notice that if characters are read into a string until EOLN or up to a nonblank delimiter character, the string may have extra leading or trailing blanks in addition to its nonblank contents. This situation will be dealt with later.

Output is perhaps straightforward enough to do without procedures. The most frequently used method will be:

```
(* Print only the part of the string in use (LEN > 0):     *)
WITH S DO
   WRITE( STR:LEN )
```

Often, however, the output is expected to use up a given number of columns, only some of which are occupied by the string:

```
(* Right-align S in N columns of output:  (N > LEN)          *)
WITH S DO
   WRITE( ' ':(N-LEN), STR:LEN );

(* Left-align S in N columns:  (N > LEN)                     *)
WITH S DO
   WRITE( STR:LEN, ' ':(N-LEN) );

(* Center S in N columns:  (N > LEN)                         *)
WITH S DO
   WRITE( ' ':((N-LEN) DIV 2), STR:LEN,
          ' ':(N - (N-LEN) DIV 2 - LEN ) );
```

Ex. 3: Write a procedure to output strings. It is to be used like this:

```
PUTSTRING( S, PLACE, N )
```

where PLACE indicates one of the three ways to position S in the N columns. Have it do something reasonable if N is less than or equal to S.LEN.

SEARCHES AND COMPARISONS

As soon as we start using strings in a program, we want to look through them to see what is in them, and to pick them apart to use the pieces. This section shows how to search for strings within strings, and how to compare strings in typical ways.

Like any array, a string can be searched to find the position of a given value, in this case a character value. More interestingly, we often search for the first occurrence of a character that belongs to a given class or set of values. Given this type:

```
TYPE                        (* Holds the targets of a search: *)
   CHARSET = SET OF CHAR;
```

then the index or position of the first letter in a string, for example, can be found this way:

```
I := INDEX( S, ['a'..'z','A'..'Z'] )
```

using this definition of the INDEX function:

```
FUNCTION  INDEX( S : STRING; VALIDS : CHARSET ) : INTEGER;
   (* Find and return the location of the first instance  *)
   (* in S of any character in the given set VALIDS:      *)
   (* (Return 0 if none is found.)                        *)

   VAR
      I : INTEGER;          (* Index for a loop.                *)
      LOC : 0..MAXLEN;      (* Location of a valid character. *)

   BEGIN (* INDEX *)

   (* This is just a variation of a linear search:          *)
   LOC := 0;
   I := 1;
   WHILE (I <= S.LEN) AND (LOC = 0) DO
      IF S.STR[I] IN VALIDS THEN
         LOC := I
      ELSE
         I := I + 1;
   INDEX := LOC

   END; (* INDEX *)
```

†*Ex. 4: Write a function that returns the index of the first character that is not in a given set of values, or 0 if all are in the set. Why would you want such a function?*

A common search is to look for a given *word* inside a string. More generally, we often want to know if and where *any* given string is inside another given string:

```
FUNCTION  POSITION( OBJ : STRING; SUBJ : STRING ) : INTEGER;
   (* Return the position (index of the first character)  *)
   (* of OBJ in the string SUBJ.  If OBJ does not occur   *)
   (* as a substring of SUBJ, return a 0:                 *)

   VAR
      START : INTEGER;   (* Position in SUBJ to start with.*)
      FOUND : BOOLEAN;   (* TRUE when OBJ is in SUBJ.      *)
      MATCH : BOOLEAN;   (* TRUE if OBJ is matching SUBJ.  *)
      I : INTEGER;       (* Loop index.                    *)

   BEGIN (* POSITION *)

   (* If either string is empty, the position is 0:       *)
   IF (OBJ.LEN = 0) OR (SUBJ.LEN = 0) THEN
      POSITION := 0
   ELSE
      BEGIN (* searches *)
      START := 1;
      FOUND := FALSE;

      (* Search SUBJ for the first character of OBJ, but  *)
      (* only when there is room enough in SUBJ for the   *)
      (* rest of OBJ starting at the given point (START): *)
      WHILE ((SUBJ.LEN-START+1) >= OBJ.LEN) AND NOT FOUND DO
         BEGIN (* find first *)
         IF SUBJ.STR[START] = OBJ.STR[1] THEN
            BEGIN (* compare rest *)
            MATCH := TRUE;
            I := 1;
            WHILE (I < OBJ.LEN) AND MATCH DO
               IF SUBJ.STR[START+I] = OBJ.STR[1+I] THEN
                  I := I + 1
               ELSE
                  MATCH := FALSE;
            FOUND := MATCH
            END (* compare rest *)
         ELSE
            START := START + 1
         END; (* find first *)

      IF FOUND THEN
         POSITION := START
      ELSE
         POSITION := 0
      END (* searches *)

   END; (* POSITION *)
```

Ex. 5: Write a procedure called NEXTWORD that will return the next word encountered in a given string starting at a given position in that string. (Define "word" as any string of nonblanks.)

Ex. 6: Modify NEXTWORD so that it also tells where in the given string it encountered the word.

Ex. 7: Use NEXTWORD to find and print all the words in a given sentence, one per line.

It would seem that string comparisons are trivial, but this is true only if blanks are not considered significant to the comparison. In this case, the two STR fields could be compared. But if the program requires that two blanks not be considered equal to three, or that, for example, 'able' is not the same as 'able ', then you might use this function:

```
FUNCTION  EQUAL( A : STRING; B : STRING ) : BOOLEAN;
   (* Return TRUE if the strings A and B are the same       *)
   (* length and have the same content:                     *)

   BEGIN (* EQUAL *)

   EQUAL := (A.LEN = B.LEN) AND (A.STR = B.STR)

   END; (* EQUAL *)
```

Ex. 8: Why do we insist that the unused portion of a string be kept blank?

Even if we were not concerned with the lengths of the portions of the strings "in use" when they are compared, we might still use a comparison function when there are many comparisons to do, or just to hide the details of the strings:

```
FUNCTION  COMPARE( A : STRING; B : STRING ) : INTEGER;
   (* Return -1 if A < B, 0 if A = B, or 1 if A > B.        *)

   BEGIN (* COMPARE *)

   IF A.STR < B.STR THEN
      COMPARE := -1
   ELSE IF A.STR > B.STR THEN
      COMPARE := 1
   ELSE
      COMPARE := 0

   END; (* COMPARE *)
```

This function could be used with a case statement:

```
CASE COMPARE( S1, S2 ) OF
   -1: ...;
    0: ...;
    1: ...
   END (* CASE *)
```

where it might be better to return a value of an enumerated type (LESS, SAME, MORE). It could also be used with a test against 0, where the comparison operator indicates the relationship of the strings:

```
IF COMPARE( S1, S2 ) < 0 THEN
   (* S1 < S2 *)...
```

†*Ex. 9: Change COMPARE to account for the lengths of the strings. Assume an ordering wherein*

```
'a' < 'a ' < 'b' < 'b '
```

and so on.

EDITING

This section shows how to pick strings apart to use the pieces. We need these algorithms because strings almost always represent words or phrases, program statements, algebraic formulae, or some other entities composed of smaller strings that are interesting in themselves. We might want to isolate those smaller strings and manipulate them. On the other hand, given the pieces, we often want to assemble them into larger strings. Algorithms for this are given first.

Strings are often assembled by accumulating one character at a time on the righthand end. This is called *catenation.* The characters to be catenated may come from input or from another string. One algorithm for catenation is:

```
PROCEDURE  CATCHAR( VAR S : STRING; CH : CHAR );
   (* Catenate a given char CH to a given string S, if    *)
   (* there is room for it:                               *)

   BEGIN (* CATCHAR *)

   WITH S DO
      IF LEN < MAXLEN THEN
         BEGIN (* append *)
         LEN := LEN + 1;
         STR[LEN] := CH
         END (* append *)

   END; (* CATCHAR *)
```

Ex. 10: Use the CATCHAR procedure in a loop that inputs characters and saves them into S.

†*Ex. 11: What are two advantages and one disadvantage of using CATCHAR in the input loop of the previous exercise?*

We will also want to catenate entire strings. For example, words might be catenated to create sentences. An algorithm for this is:

```
PROCEDURE  CATSTRING( VAR OBJ : STRING; S : STRING );
   (* Catenate a given string S to the given string OBJ.  *)
   (* If the combination is too long for OBJ, the extra   *)
   (* characters on the right are discarded.              *)

   VAR
      I : INTEGER;        (* Loop index.                  *)
      AMOUNT : INTEGER; (* Number of S's chars to use.    *)

   BEGIN (* CATSTRING *)

   (* Find how much of S will fit on the end of OBJ:      *)
   IF OBJ.LEN+S.LEN > MAXLEN THEN
      AMOUNT := MAXLEN - OBJ.LEN
   ELSE
      AMOUNT := S.LEN;

   (* Copy characters from S to the end of OBJ:           *)
   WITH OBJ DO
      FOR I:=1 TO AMOUNT DO
         BEGIN (* catenate *)
         LEN := LEN + 1;
         STR[LEN] := S.STR[I]
         END (* catenate *)

   END; (* CATSTRING *)
```

Ex. 12: Rewrite the body of CATSTRING in terms of CATCHAR.

In both CATCHAR and CATSTRING, if the combined string would be longer than MAXLEN, then the extra characters are ignored. This is called "truncation (cutting off) on the right."

The algorithms in the next group take strings apart. Probably the most frequently used of these is an algorithm to make a copy of part of another string (which is itself left unchanged):

```
PROCEDURE  COPY( SOURCE : STRING; POS : INTEGER;
                 NUM : INTEGER; VAR DEST : STRING );
   (* Copy NUM characters from SOURCE to DEST, starting   *)
   (* at position POS in SOURCE.  If any of the indicated *)
   (* substring of SOURCE does not exist, that portion is *)
   (* considered an empty or null string (of length 0).   *)

   VAR
      I : INTEGER;          (* Loop index.                *)
      LAST : INTEGER;       (* Locus of last SOURCE char used.*)

   BEGIN (* COPY *)

   DEST := NULL;

   (* Find out how much to copy, then do it:              *)
   IF POS <= SOURCE.LEN THEN
      BEGIN (* copying *)

      (* Find the last position of substring indicated:   *)
      IF (POS + NUM - 1) >= SOURCE.LEN THEN
         LAST := SOURCE.LEN
      ELSE
         LAST := POS + NUM - 1;

      (* Copy the substring from SOURCE to DEST:          *)
      WITH DEST DO
         FOR I:=POS TO LAST DO
            BEGIN
            LEN := LEN + 1;
            STR[LEN] := SOURCE.STR[I]
            END
      END (* copying *)

   END; (* COPY *)
```

Ex. 13: Does COPY(S1,0,0,S2) work? Does COPY(S1,0,1,S2) work? Does COPY(S1,1,0,S2) work? Fix COPY to protect it from any calls that do not work (by returning an empty string). Are these calls likely to occur?

†*Ex. 14: Use COPY and CATSTRING to write an INSERT procedure that will insert one string into the middle of another, beginning at a given position, thus creating a longer string.*

After a string is found within a longer string (perhaps using POSITION) it might be removed from the longer string. This might be done to get it out of the way, or because it is no longer needed, or for some other reason. An algorithm to do this is:

```
PROCEDURE  DELETE( VAR S : STRING; POS : INTEGER;
                                   NUM : INTEGER );
   (* Delete NUM characters from S, starting at position  *)
   (* POS.  If the substring indicated by POS and NUM      *)
   (* overlaps the end of S, only the overlapping part     *)
   (* is deleted.  A substring that is not in S has no     *)
   (* effect.                                              *)

   VAR
      I : INTEGER;          (* Loop index.                 *)
      DEST : INTEGER;       (* For shifting the retained part.*)

   BEGIN (* DELETE *)

   (* See if the indicated string can be deleted:         *)
   IF POS > S.LEN THEN
      (* Do nothing--substring is beyond the end of S.    *)
   ELSE IF (POS + NUM - 1) >= S.LEN THEN
      (* Substring overlaps--get rid of overlapped part:  *)
      S.LEN := POS - 1
   ELSE
      BEGIN (* delete middle *)

      (* Shift the retained segment on the right into the *)
      (* gap left by the deleted substring:               *)
      DEST := POS;
      WITH S DO
         FOR I:=POS+NUM TO MAXLEN DO
            BEGIN (* shift *)
            STR[DEST] := STR[I];
            DEST := DEST + 1
            END; (* shift *)
      S.LEN := S.LEN - NUM
      END; (* delete middle *)

   (* In any case, blank out the unused portion of S:     *)
   FOR I:=S.LEN+1 TO MAXLEN DO
      S.STR[I] := ' '

   END; (* DELETE *)
```

†*Ex. 15: Will DELETE delete an entire string, as in DELETE(S,1,MAXLEN)?*

Ex. 16: Rewrite DELETE in terms of COPY and CATSTRING. Make sure that it will properly delete the last or first character of a string.

We noticed earlier that for one reason or another a string may contain trailing blanks—that is, blanks on the right *within* the used portion. Some algorithms require that these be removed before the strings are used. This is called "trimming:"

```
PROCEDURE  TRIM( VAR S : STRING );
   (* Trim the trailing blanks, if any, from S.            *)

   VAR
      DONE : BOOLEAN;     (* TRUE when S has been trimmed.  *)

   BEGIN (* TRIM *)

   DONE := FALSE;
   WITH S DO
      WHILE (LEN > 0) AND NOT DONE DO
         IF STR[LEN] = ' ' THEN
            LEN := LEN - 1
         ELSE
            DONE := TRUE

   END; (* TRIM *)
```

†*Ex. 17: Use TRIM to write a procedure MAKESTRING that makes a STRING type out of a given FIXEDSTRING value. (This is quite handy if MAXLEN is short enough to allow literal strings as arguments. For example:*

```
MAKESTRING( S, 'word        ' )
```

produces a STRING value in S with a LEN of 4.)

There are many, many more *possible* string-manipulation procedures, but they are used very seldom.

Perhaps this section should be ended with a lesson concerning the wisdom of using string-processing procedures and functions. Consider these exercises:

Ex. 18: Remove all the blanks from the string S (perhaps to compress S) by using INDEX to find them and DELETE to take them out.

Ex. 19: Remove the blanks from S this way: Assemble a temporary string by stepping through S and catenating its nonblanks (using CATCHAR) to the temporary. Then assign the temporary to S.

Ex. 20: Remove the blanks from S using the list-compression algorithm from Part III.

The first solution requires over 50 lines of program when the procedures are included in the count. The second requires half as many, but still more than the last version. A program that used the first algorithm would be larger and more costly to run than one that used the third algorithm. On the other hand, disregarding the procedures themselves, the first algorithm is tidy and easy to follow. The lesson is that you should carefully consider the size and expense of using string-processing procedures and weigh these drawbacks against their convenience. And, of course, string-processing should not be used when character-processing is possible.

INPUT DATA CHECKING

When a program is used by persons other than the programmer, someone is likely to present it with improper data that could cause it to yield spurious results or terminate abnormally. It is often best to write programs so that they protect themselves against such errors; a program that does so is called "robust."

Although input data checking and error handling is a broad subject with varying philosophies and implementations, the fundamental ideas are simple: the data are read as strings, and the strings are inspected to see if they represent properly formed data; if not, error messages are printed and other actions are usually taken; if correct, the strings may be converted to other (the intended) data types.

As an example, assume that several columns of a line of input are supposed to represent an integer—that is, they should all contain character digits. We already know how to read these characters into a string. The string can then be checked to see if it is indeed "numeric" using this function:

```
FUNCTION  NUMERIC( S : STRING ) : BOOLEAN;
   (* Return TRUE if and only if the characters in S      *)
   (* represent an (unsigned) integer.                     *)

   VAR
      I : INTEGER;          (* Index into S.                  *)
      OK : BOOLEAN;         (* TRUE while chars are digits.   *)

   BEGIN (* NUMERIC *)

   OK := TRUE;
   I := 1;
   WHILE (I <= S.LEN) AND OK DO
      IF S.STR[I] IN ['0'..'9'] THEN
         I := I + 1
      ELSE
         OK := FALSE;
   NUMERIC := OK

   END; (* NUMERIC *)
```

Ex. 21: Rewrite NUMERIC to allow for leading blanks and a sign.

Ex. 22: Write a REALNUMBER function that checks a string to see if it is in the form, for example, 'ddd.dd' (where d is a digit).

If the string does represent an integer, it may have to be converted to an actual integer so that it can be used in computations. (An ID number, on the other hand, could just as well be a string.) A function to do the conversion could look like this:

```
FUNCTION  INTVALUE( S : STRING ) : INTEGER;
   (* Convert a string of character digits into the        *)
   (* corresponding integer value.                          *)

   VAR
      NUMBER : INTEGER; (* Holds the integer value.         *)
      I : INTEGER;      (* Index into S.                    *)

   BEGIN (* INTVALUE *)

   NUMBER := 0;
   FOR I:=1 TO S.LEN DO
      NUMBER := NUMBER*10 + (ORD(S.STR[I])-ORD('0'));
   INTVALUE := NUMBER

   END; (* INTVALUE *)
```

Ex. 23: Rewrite INTVALUE to allow for leading blanks and a sign. (What about an entirely blank string?)

†*Ex. 24: Write a REALVALUE function to convert strings that are in decimal form.*

The two functions might then be used in combination this way:

```
IF NUMERIC( S ) THEN
   N := INTVALUE( S )
ELSE
   WITH S DO
      WRITELN( '***Error: ', STR:LEN, ' is not numeric.' )
```

TEXT FORMATTING

The text of this book was formatted with a text processor. Text (or word) processors take, as their input, text interspersed with formatting "commands." The commands are special characters or words that tell the program how to arrange the text for output—where to break it into pages, paragraphs, or lines, how wide the lines should be, which lines to center or underscore, and so on. The program for reading at the end of this part is a very simple text processor. Like all such programs, it makes heavy use of string algorithms.

To a text processor, the input is a sequence of words that are accumulated into lines (not sentences) and then arranged for output. The lines are just strings of a given length. The words are "arranged" by various algorithms. For example:

Left-alignment: some words are shifted as a group to the left end of the line.

Right-alignment: some words are shifted to the right end.

Centering: some words are centered within a line.

Justification: all of the words on a line are spread out so that they are distributed from the left end to the right end, as in most books.

The example program uses the algorithms, strategies, and control structures that are typical of text processors.

PROGRAM FOR READING

```
 1  PROGRAM  FORMATTER( INPUT, OUTPUT );
 2
 3  (* PURPOSE:                                                        *)
 4  (*    A simple general purpose text formatter that uses            *)
 5  (*    embedded formatting commands.                                *)
 6  (*                                                                 *)
 7  (* PROGRAMMER:   David V. Moffat                                   *)
 8  (*                                                                 *)
 9  (* INPUT & OUTPUT:                                                 *)
10  (*    The input is any text containing any of the special         *)
11  (*    characters that are used as formatting commands.            *)
12  (*    The output is a new version of the given text,              *)
13  (*    without the commands, but formatted as the commands         *)
14  (*    indicated.  The commands, their names, and their            *)
15  (*    effects are:                                                 *)
16  (*    CMND NAME      EFFECT                                        *)
17  (*     /   NEWLINE* Print the current line, start a new           *)
18  (*                  line.                                          *)
19  (*     %   NEWPAGE* Print current line, start a new line          *)
20  (*                  on a new page.                                 *)
21  (*     <   LEFT*    Left align all text since the last            *)
22  (*                  "action" (names with "*" are actions).        *)
23  (*     =   CENTER*  Center all text since the last action.        *)
24  (*     >   RIGHT*   Right align all text since the last           *)
25  (*                  action, print it and start new line.          *)
26  (*     !   LITERAL  Print the next character literally.           *)
27  (*                  If it is a command, do not execute it.        *)
28  (*     @   ASIS     Retain all characters after this,             *)
29  (*                  regardless of what they are, up to           *)
30  (*                  the next "@" or end of the input line.        *)
31  (*                                                                 *)
32  (*   The output line is 65 columns.  Words of input are          *)
33  (*   accumulated until no more will fit on the current           *)
34  (*   line of output.  The line is then justified and             *)
35  (*   printed, and a new line begun.  This continues              *)
36  (*   until end-of-file or unless temporarily interrupted         *)
37  (*   by any of the commands.  Usually, one blank is kept         *)
38  (*   between words, regardless of the input, until the           *)
39  (*   line is justified.  More blanks can be specifically         *)
40  (*   kept (called "significant blanks") using "!" or "@".        *)
41  (*                                                                 *)
42  (* ASSUMPTIONS/LIMITATIONS:                                        *)
43  (*   Meaningless combinations of commands are not found.         *)
44
```

```
45  (* ALGORITHM/STRATEGY:                                           *)
46  (*    This program uses the state-transition strategy to        *)
47  (*    process the input characters, accumulating them           *)
48  (*    into words and lines, and reacting to the commands.       *)
49  (*                                                              *)
50  (*    Initialize.                                               *)
51  (*    Process input characters until end-of-file:               *)
52  (*       Get a character and its class.                         *)
53  (*       Use class and current state to select action.          *)
54  (*    Finish up last line.                                      *)
55
56  CONST
57     MAXLEN = 80;               (* Max length of any string.  *)
58     PRINTWIDTH = 65;           (* Max columns on output line.*)
59     MARGIN = '          ';     (* 10-column line indention.  *)
60
61  TYPE
62     FIXEDSTRING = PACKED ARRAY[1..MAXLEN] OF CHAR;
63
64     STRING =                   (* Varying length strings:    *)
65        RECORD
66        STR : FIXEDSTRING;      (* Text of string.            *)
67        LEN : 0..MAXLEN         (* Amount of string in use.   *)
68        END;
69
70     OUTPUTLINE =               (* Line of accumulated words. *)
71        RECORD
72        OUT : STRING;           (* Actual line to be printed. *)
73        ACTIONCOL : INTEGER     (* Position of last action:   *)
74        END;                    (* last affected column + 1.  *)
75
76     CHARCLASSES =              (* Char processing classes:   *)
77       (BLANK,                  (* Blanks only.               *)
78        NEWLINE,                (* Command characters:  "/"   *)
79        NEWPAGE,                (*                      "%"   *)
80        ASIS,                   (*                      "@"   *)
81        LEFT,                   (*                      "<"   *)
82        CENTER,                 (*                      "="   *)
83        RIGHT,                  (*                      ">"   *)
84        LITERAL,                (*                      "!"   *)
85        NONBLANK,               (* All other nonblanks.       *)
86        ENDLINE,                (* EOLN(INPUT).               *)
87        ENDFILE);               (* EOF(INPUT).                *)
88
89     STATES =                   (* Processing states:         *)
90       (NORMAL,                 (* Handling text and commands.*)
91        ASISMODE,               (* Printing all characters.   *)
92        LITERALNEXT);           (* Waiting to print next char.*)
93
```

```
 94   VAR
 95      LINE : OUTPUTLINE;         (* Line of words to print.    *)
 96      WORD : STRING;             (* Current word of input.     *)
 97      CH : CHAR;                 (* One character of input.    *)
 98      CLASS : CHARCLASSES;       (* Input character's class.   *)
 99      STATE : STATES;            (* Current processing state.  *)
100      NULL : STRING;             (* Empty string to init others*)
101      SIGBLANK : CHAR;           (* Set to CHR(0), the         *)
102                                 (* "significant blank" code.  *)
103      CLASSOF : ARRAY[CHAR] OF CHARCLASSES;
104                                 (* "Function" to classify CH. *)
105
106
107
108   PROCEDURE  INITIALIZE;
109      (* Initialize the global variables CLASSOF, NULL, and    *)
110      (* SIGBLANK.                                             *)
111
112      VAR
113         CH : CHAR;              (* Index into CLASSOF.        *)
114         I : 1..MAXLEN;          (* Index into NULL.STR.       *)
115
116      BEGIN (* INITIALIZE *)
117
118      (* Set up "function" array to classify characters:      *)
119      FOR CH:=CHR(0) TO CHR(253) DO (* Waterloo reserves 254 *)
120         CLASSOF[CH] := NONBLANK;
121      CLASSOF['/'] := NEWLINE;
122      CLASSOF['%'] := NEWPAGE;
123      CLASSOF['@'] := ASIS;
124      CLASSOF['>'] := RIGHT;
125      CLASSOF['='] := CENTER;
126      CLASSOF['<'] := LEFT;
127      CLASSOF['!'] := LITERAL;
128      CLASSOF[' '] := BLANK;
129
130      (* Initialize the "empty" string for clearing others:   *)
131      FOR I:=1 TO MAXLEN DO
132         NULL.STR[I] := ' ';
133      NULL.LEN := 0;
134
135      (* Set up a "significant blank":                        *)
136      SIGBLANK := CHR(0)
137
138      END; (* INITIALIZE *)
139
```

```
140   PROCEDURE GETNEXT( VAR CH : CHAR; VAR CLASS : CHARCLASSES );
141      (* Get the next char, CH, of input and find its CLASS. *)
142      (* Return a blank in CH for EOLN and EOF.               *)
143
144      BEGIN (* GETNEXT *)
145
146      IF EOF(INPUT) THEN
147         BEGIN
148         CLASS := ENDFILE;
149         CH := ' '
150         END
151      ELSE IF EOLN(INPUT) THEN
152         BEGIN
153         CLASS := ENDLINE;
154         READ( CH )           (* Reads as a blank.            *)
155         END
156      ELSE
157         BEGIN
158         READ( CH );
159         CLASS := CLASSOF[CH]
160         END;
161
162      END; (* GETNEXT *)
163
164
165
166   PROCEDURE  CATCHAR( VAR S : STRING; CH : CHAR );
167      (* Catenate CH to S if there is room.                   *)
168
169      BEGIN (* CATCHAR *)
170
171      WITH S DO
172         IF LEN < MAXLEN THEN
173            BEGIN
174            LEN := LEN + 1;
175            STR[LEN] := CH
176            END
177
178      END; (* CATCHAR *)
179
180
```

```
181   PROCEDURE  CATSTRING( VAR S : STRING; NEW : STRING );
182      (* Catenate NEW to the right end of S.                        *)
183
184      VAR
185         I : 1..MAXLEN;          (* Loop index.                     *)
186
187      BEGIN (* CATSTRING *)
188
189      WITH S DO
190         FOR I:=1 TO NEW.LEN DO
191            BEGIN
192            LEN := LEN + 1;
193            STR[LEN] := NEW.STR[I]
194            END
195
196      END; (* CATSTRING *)
197
198
199
200   PROCEDURE  SHIFT( VAR LINE : OUTPUTLINE; FINAL : INTEGER );
201      (* Shift the substring of the LINE (starting at             *)
202      (* the position of last action and ending at the            *)
203      (* current length) to right align it at position FINAL.*)
204
205      VAR
206         I : 1..MAXLEN;          (* Loop index.                     *)
207         DEST : INTEGER;         (* Index for moving chars.         *)
208
209      BEGIN (* SHIFT *)
210
211      (* Shift the chars, and blank out their old positions: *)
212      WITH LINE, OUT DO
213         IF (ACTIONCOL <= LEN) AND (LEN < PRINTWIDTH) AND
214            (LEN < FINAL) THEN
215            BEGIN
216            DEST := FINAL;
217            FOR I:=LEN DOWNTO ACTIONCOL DO
218               BEGIN (* move each *)
219               STR[DEST] := STR[I];
220               STR[I] := ' ';
221               DEST := DEST - 1
222               END (* move each *)
223            END;
224      (* This is now the point of last action:                  *)
225      WITH LINE, OUT DO
226         BEGIN
227         LEN := FINAL;
228         ACTIONCOL := LEN + 1
229         END
230      END; (* SHIFT *)
231
```

```
232  PROCEDURE  ALIGN( VAR LINE : OUTPUTLINE;
233                    DIRECTION : CHARCLASSES );
234     (* Adjust a group of words on LINE so that they are    *)
235     (* aligned as indicated by DIRECTION.                   *)
236
237     VAR
238        NEWLENGTH : INTEGER;  (* Length after alignment.    *)
239
240     BEGIN (* ALIGN *)
241
242     (* Determine the new length of the line:                *)
243     WITH LINE, OUT DO
244        CASE DIRECTION OF
245           LEFT:
246              (* Length now is alignment point, by defn.:   *)
247              NEWLENGTH := LEN;
248           CENTER:
249              (* Length becomes half the word group's       *)
250              (* length plus half the maximum line length:  *)
251              NEWLENGTH := ((LEN-ACTIONCOL+1) DIV 2) +
252                           (PRINTWIDTH DIV 2);
253           RIGHT:
254              (* The words move all the way to the right:   *)
255              NEWLENGTH := PRINTWIDTH
256        END; (* DIRECTION cases *)
257     (* Have the substring containing the words aligned:    *)
258     SHIFT( LINE, NEWLENGTH )
259
260     END; (* ALIGN *)
261
```

```
262   PROCEDURE  JUSTIFY( VAR LINE : OUTPUTLINE );
263      (* Justify LINE to a length of PRINTWIDTH by putting    *)
264      (* extra blanks between words, from right to left.      *)
265      (* The line currently has one blank between words.      *)
266
267      VAR
268         BLANKS : INTEGER;        (* # of blanks to be inserted.*)
269         GAPS : INTEGER;          (* # of gaps between words.   *)
270         N : INTEGER;             (* Amount to expand one gap.  *)
271         DEST : INTEGER;          (* New place for a moved char.*)
272         SOURCE : INTEGER;        (* Source column of that char.*)
273
274      BEGIN (* JUSTIFY *)
275
276      WITH LINE, OUT DO
277         (* Spread the line out if it is too short:              *)
278         IF LEN < PRINTWIDTH THEN
279            BEGIN
280            (* Count the number of gaps between words:          *)
281            GAPS := 0;
282            FOR SOURCE:=1 TO LEN DO
283               IF STR[SOURCE] = ' ' THEN
284                  GAPS := GAPS + 1;
285            (* Find # of blanks needed to stretch the line:   *)
286            BLANKS := PRINTWIDTH - LEN;
287            (* Shift characters to the right, distributing    *)
288            (* extra blanks between the words (in the gaps): *)
289            DEST := PRINTWIDTH;
290            SOURCE := LEN;
291            WHILE GAPS > 0 DO
292               BEGIN (* expand line *)
293               IF STR[SOURCE] <> ' ' THEN
294                  BEGIN (* shift char *)
295                  (* Move the character and leave a blank:    *)
296                  STR[DEST] := STR[SOURCE];
297                  STR[SOURCE] := ' '
298                  END (* shift char *)
299               ELSE
300                  BEGIN (* leave blanks *)
301                  (* Find the number of blanks for this gap, *)
302                  (* and skip that many (now blank) columns: *)
303                  N := BLANKS DIV GAPS;
304                  DEST := DEST - N;
305                  GAPS := GAPS - 1;
306                  BLANKS := BLANKS - N
307                  END; (* leave blanks *)
308               (* Step to next source and destination chars: *)
309               SOURCE := SOURCE - 1;
310               DEST := DEST - 1
311               END; (* expand line *)
312            LEN := PRINTWIDTH
313            END
314      END; (* JUSTIFY *)
```

```
315
316
317  PROCEDURE  PRINT( S : STRING );
318     (* Print a left margin and the string, S, (changing     *)
319     (* the "significant blanks" to blanks) on one line.     *)
320
321     VAR
322        I : 1..MAXLEN;        (* For stepping through S.     *)
323
324     BEGIN (* PRINT *)
325
326     WRITE( MARGIN );
327     WITH S DO
328        FOR I:=1 TO LEN DO
329           IF STR[I] = SIGBLANK THEN
330              WRITE( ' ' )
331           ELSE
332              WRITE( STR[I] );
333     WRITELN
334
335     END; (* PRINT *)
336
337
338
339  PROCEDURE  START( VAR LINE : OUTPUTLINE );
340     (* Initialize the given LINE to empty.                  *)
341
342     BEGIN (* START *)
343
344     WITH LINE DO
345        BEGIN
346        OUT := NULL;
347        ACTIONCOL := 1
348        END
349
350     END; (* START *)
351
352
353
354  PROCEDURE  FINISH( VAR LINE : OUTPUTLINE );
355     (* Print this LINE and reinitialize it.                 *)
356
357     BEGIN (* FINISH *)
358
359     PRINT( LINE.OUT );
360     START( LINE )
361
362     END; (* FINISH *)
363
```

```
364   PROCEDURE  SAVE( VAR LINE : OUTPUTLINE; VAR WORD : STRING );
365      (* Put the WORD onto the end of the LINE, if possible. *)
366      (* Otherwise finish the current LINE and try again.    *)
367      (* Reinitialize the WORD to empty.                      *)
368
369      BEGIN (* SAVE *)
370
371      (* Try to put the word (preceded by a blank if it is   *)
372      (* not the first word) on the end of the line:          *)
373      IF WORD.LEN = 0 THEN
374         (* Called due to extra blanks in input--do nothing. *)
375      ELSE IF (LINE.OUT.LEN + WORD.LEN + 1) <= PRINTWIDTH THEN
376         BEGIN (* it fits *)
377         (* The blank and the word will fit:                  *)
378         IF LINE.OUT.LEN > 0 THEN
379            CATCHAR( LINE.OUT, ' ' );
380         CATSTRING( LINE.OUT, WORD )
381         END (* it fits *)
382      ELSE
383         (* The combination is too long.  See if it is just  *)
384         (* the word or everything together:                  *)
385         IF WORD.LEN >= PRINTWIDTH THEN
386            BEGIN (* oversized word *)
387            (* Print the line, then force the word out:       *)
388            JUSTIFY( LINE );
389            FINISH( LINE );
390            PRINT( WORD )
391            END (* oversized word *)
392         ELSE
393            BEGIN (* normal overflow *)
394            JUSTIFY( LINE );
395            FINISH( LINE );
396            (* Start a new line with the leftover word:       *)
397            CATSTRING( LINE.OUT, WORD )
398            END; (* normal overflow *)
399      WORD := NULL
400
401      END; (* SAVE *)
402
```

```
403  BEGIN (* FORMATTER *)
404  INITIALIZE;
405  PAGE( OUTPUT );
406  START( LINE );
407  WORD := NULL;
408  STATE := NORMAL;
409  (* Process the text one char at a time until end of file: *)
410  REPEAT
411     (* Get a character and its class:                      *)
412     GETNEXT( CH, CLASS );
413     (* Determine the action to take given the state:       *)
414     CASE STATE OF
415        NORMAL:
416           CASE CLASS OF
417              NONBLANK:
418                 (* Accumulate characters into words:        *)
419                 CATCHAR( WORD, CH );
420              BLANK, ENDLINE, ENDFILE:
421                 (* Accumulate words into print lines:       *)
422                 SAVE( LINE, WORD );
423              NEWLINE:
424                 BEGIN
425                 (* Note that "action" commands end words:   *)
426                 SAVE( LINE, WORD );
427                 FINISH( LINE )
428                 END;
429              NEWPAGE:
430                 BEGIN
431                 SAVE( LINE, WORD );
432                 FINISH( LINE );
433                 PAGE( OUTPUT )
434                 END;
435              LEFT:
436                 BEGIN
437                 SAVE( LINE, WORD );
438                 ALIGN( LINE, LEFT )
439                 END;
440              CENTER:
441                 BEGIN
442                 SAVE( LINE, WORD );
443                 ALIGN( LINE, CENTER )
444                 END;
445              RIGHT:
446                 BEGIN
447                 SAVE( LINE, WORD );
448                 ALIGN( LINE, RIGHT );
449                 FINISH( LINE )
450                 END;
451              LITERAL:
452                 STATE := LITERALNEXT;
453              ASIS:
454                 STATE := ASISMODE
455           END; (* CLASS cases *)
```

```
456
457
458
459
460        LITERALNEXT:
461           BEGIN
462           (* Take the next char literally, then go back:    *)
463           IF CLASS = BLANK THEN
464              CATCHAR( WORD, SIGBLANK )
465           ELSE
466              CATCHAR( WORD, CH );
467           STATE := NORMAL
468           END;
469
470        ASISMODE:
471           BEGIN
472           (* Take all chars literally, up to a sentinel:    *)
473           IF CLASS = BLANK THEN
474              CATCHAR( WORD, SIGBLANK )
475           ELSE IF CLASS <> ASIS THEN
476              CATCHAR( WORD, CH );
477           IF CLASS IN [ASIS,ENDLINE,ENDFILE] THEN
478              STATE := NORMAL
479           END
480     END (* STATE cases *)
481
482  UNTIL CLASS = ENDFILE;
483
484  (* Finish up the last line:                               *)
485  SAVE( LINE, WORD );
486  FINISH( LINE );
487  PAGE( OUTPUT )
488
489  END. (* FORMATTER *)
```

Example input:

```
Formatter Program<Test Run>
 //Example Output from the Text Formatting Program=//
Introduction//
@   @The normal "mode" of processing just
collects words into lines and justifies the lines by
inserting blanks between the words to spread them out so
that they are        aligned on both the right and left.
 //Special Effects//
@   @Sometimes the default justification must be suppressed
so that, for example, tables can be formed:
/@    Command  Name
/@       /     NEWLINE
/@       %     NEWPAGE
/This is why "asis mode" is necessary.  In this formatter
the "!@" starts the asis mode.  (The previous sentence
also shows why the "!!" is necessary.)  Asis mode also
allows other effects:
 /*=/* *=/@*   *@=/@*     *@=/and so on.
```

Corresponding output:

```
Formatter Program                                     Test Run


        Example Output from the Text Formatting Program

Introduction

   The normal "mode" of processing just collects words into lines
and  justifies the lines by inserting blanks between the words to
spread  them  out  so that they are aligned on both the right and
left.

Special Effects

   Sometimes  the  default  justification  must  be suppressed so
that, for example, tables can be formed:
    Command  Name
       /     NEWLINE
       %     NEWPAGE
This  is  why "asis mode" is necessary. In this formatter the "@"
starts  the  asis mode. (The previous sentence also shows why the
"!" is necessary.) Asis mode also allows other effects:
                               *
                              * *
                             *   *
                            *     *
and so on.
```

PROGRAM EXERCISES

1. Why is it that the CATSTRING procedure in the FORMATTER program can ignore the possibility of truncation?

2. Show all the changes necessary to FORMATTER to implement a paragraph command, say "|", that prints the current line and begins a new line with three significant blanks so that the text of the line will be indented.

3. Devise a state called NUMERICS that is entered from the NORMAL state when "#" is encountered. This state collects digits to form an integer. It reverts to the NORMAL state when it encounters a nondigit—but first uses the integer to set a new margin width, print width, or spacing between output lines, depending upon whether the nondigit is "M", "P", or "S", respectively (all others are ignored). Thus, "#50P" in the input sets the printing width to 50 columns from that point on. (It is currently determined by a constant.)

4. Show how and where input data checking can be put into the CHECKING program of Part IV. What are the appropriate reactions to input errors *after* they have been found and reported?

5. How can the FORMATTER program be changed to allow the program user to decide which characters to use as formatting commands? (Be sure to consider the user documentation.)

Part VI: File Algorithms

INTRODUCTION

The standard files INPUT and OUTPUT have been used throughout this book for getting information into the various data structures in the programs and for displaying results. This part focuses upon files as generalized data structures in their own right.

Like one-dimensional arrays, files are linear and homogeneous (all the elements or components of a given file are of the same type). Unlike array access, which can be to any item at any time (random access), file access is purely sequential; items can be accessed only in the order in which they are stored. Unlike arrays, which must have declared sizes or bounds, files are unbounded; they have a necessary beginning but no predetermined end. Because of these properties, a file is often called a *sequence.*

Assume that we have these declarations:

```
TYPE
   INFOTYPE = ...;
   KEYTYPE = ...;
   ITEM =                   (* Each file element:            *)
      RECORD
      KEY : KEYTYPE;        (* "Key" to the data. (See text.) *)
      INFO : INFOTYPE       (* Information to be stored.      *)
      END;
   ITEMFILE = FILE OF ITEM;
                            (* A "sequence" of these ITEMs.   *)

VAR
   F : ITEMFILE;            (* A file, and a variable whose   *)
   I : ITEM;                (* value can be put into the file.*)
```

Files are most frequently used to store sequences of structured variables such as records. As was shown in the discussion of tables in Part IV, structured values often include a ''key'' value, which somehow represents the other values in the structure. For example, license numbers may be the keys to records containing drivers' names, addresses, and so on.

The file or sequence F is initialized to empty and prepared to receive information (readied for output) using the predefined procedure REWRITE(F). Files can be illustrated this way:

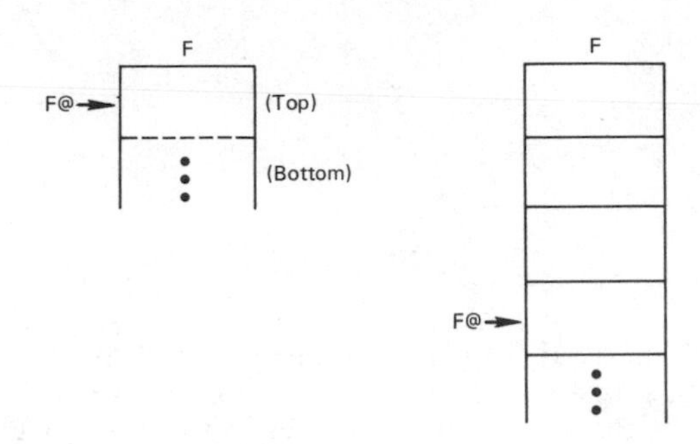

The first picture shows the top, the open-ended bottom, and the so-called "buffer variable" or "file pointer," F@, which can be thought of as pointing to the *next available location* in the sequence into which the value of an item can be placed. The second picture shows the sequence after three values have been put into it, using, for example, WRITE(F,I) three times. (The "@" is a synonym for the caret in standard Pascal. It is easier to see and is more mnemonic!)

The values in the sequence F can be inspected only after the procedure RESET(F) is applied to it, which prepares the file for inspection (readies it for input), yielding this situation:

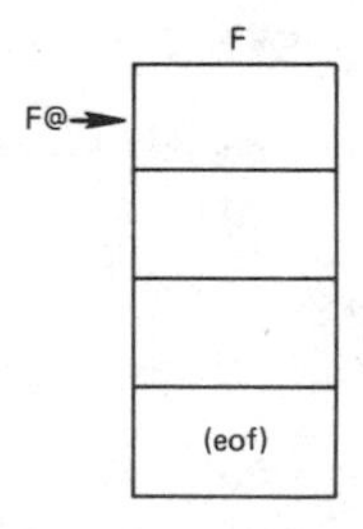

Now the file pointer points to the first value of type ITEM that can be referenced, using READ(F,I), for example. It always points to the *next available value* in the file. The sequence of values is terminated by a system-supplied sentinel; the function EOF(F) returns TRUE if and only if the file pointer is positioned at that sentinel.

Note that the point of reference for the terms "input" and "output" is the *program.* Thus, "output" means to copy information *from* the program *into* a file!

It is important to remember that F@ acts much like a variable of type ITEM, and that the file pointer can be moved along the sequence using GET(F) or PUT(F) if F is ready for input or output, respectively. In particular, recall that:

```
WRITE( F, I )
```

is the same as

```
BEGIN
(* Place the value into the next file location:                *)
F@ := I;
(* Move the pointer to the next file location:                 *)
PUT( F )
END
```

while

```
READ( F, I )
```

is the same as

```
BEGIN
(* Copy the value from this file location:                     *)
I := F@;
(* Move the pointer to the next file location:                 *)
GET( F )
END
```

The file pointers and the GET and PUT procedures are often used separately for efficiency or to simplify algorithms.

The standard files INPUT and OUTPUT are of type TEXT, which is a predefined name for FILE OF CHAR. All files of type TEXT have all the properties just discussed; their elements (or components) are single characters. In addition, they have several special features that are not shared by other files. These special features were designed into the standard files for extra convenience (they are the most frequently used files):

End-of-line characters to terminate "logical" lines, an EOLN function to detect them, and READLN and WRITELN procedures to manipulate them.

The PAGE procedure.

The ability to input or output the values of several variables at once.

Automatic conversion of values between their actual types and their representations as legible characters in the files.

Finally, note that there are many ways in which the various implementations of Pascal require the programmer to make the necessary connections (if any) between the files used in a program and the actual data given to the program by the operating system. You should consult the documentation for the Pascal compiler that you use.

BASIC OPERATIONS

The basic operations that are performed upon lists and tables are also performed upon files, because lists, tables, and files are all homogeneous linear data structures.

Files are usually created using initial data that are read in from standard input. Assuming one item per line of input:

```
PROCEDURE  INITFILE( VAR F : ITEMFILE );
   (* Initialize file F using data from INPUT.              *)

   VAR
      I : ITEM;           (* For reading and writing values.*)

   BEGIN (* INITFILE *)

   (* Empty the file and prepare it for output:              *)
   REWRITE( F );

   (* Copy data from file INPUT (which is always ready       *)
   (* for input) to F:                                       *)
   WHILE NOT EOF(INPUT) DO
      BEGIN (* each item *)
      READLN( INPUT, I.KEY, I.INFO );
      WRITE( F, I )
      END (* each item *)

   END; (* INITFILE *)
```

The formal, rather than the simplified, notation for standard I/O is used here so that the files used are clearly distinguished. Note that the KEY values could have been generated by the procedure itself or elsewhere in the program, rather than be part of the original input data. Also recall that the values in the INPUT file are represented as *characters* and are automatically converted into numeric values when necessary, while the values put into F are unchanged when they are copied from the program into the file.

The general algorithm for copying values from one file to another does not need extra variables, since it can manipulate the file pointers directly:

```
PROCEDURE  COPYFILE( VAR OLD: ITEMFILE; VAR NEW: ITEMFILE );
   ( * Copy items from file OLD into NEW.                     *)

   BEGIN (* COPYFILE *)

   (* Prepare files for I/O:                                  *)
   RESET( OLD );
   REWRITE( NEW );

   (* Copy the contents directly, using file pointers:        *)
   WHILE NOT EOF(OLD) DO
      BEGIN (* each item *)
      NEW@ := OLD@;
      PUT( NEW );
      GET( OLD )
      END (* each item *)

   END; (* COPYFILE *)
```

Ex. 1: Write the simple, yet often used, procedure called COPYREST, which merely copies the first of two given files to the second, assuming that they are already prepared for I/O.

The algorithm for displaying a file (printing its contents to the standard file OUTPUT) is similar:

```
PROCEDURE  PRINTFILE( VAR F : ITEMFILE );
   (* Print the contents of F in tabular form.                *)

   BEGIN (* PRINTFILE *)

   (* Prepare file F for input (OUTPUT is always ready):      *)
   RESET( F );
   WRITELN( OUTPUT, 'Key':10, 'Information':15 );

   (* Copy the contents of F to OUTPUT:                       *)
   WHILE NOT EOF(F) DO
      BEGIN (* each item *)
      WRITELN( OUTPUT, F@.KEY:10, F@.INFO:15 );
      GET( F )
      END (* each item *)

   END; (* PRINTFILE *)
```

Rather than print out the contents of an entire file, we often want only to retrieve the information that corresponds to a given KEY value. The file must be searched to find the value:

```
PROCEDURE  RETRIEVE( VAR F : ITEMFILE; VAR I : ITEM );
   (* Find the item in F whose KEY value equals I.KEY,     *)
   (* and return the rest of that item via I.  (The given *)
   (* key value is assumed to exist somewhere in F.)      *)

   BEGIN (* RETRIEVE *)

   (* Ready the file, and search for the given KEY value: *)
   RESET( F );
   WHILE F@.KEY <> I.KEY DO
      GET( F );

   (* Return the other information:                        *)
   I.INFO := F@.INFO

   END; (* RETRIEVE *)
```

†*Ex. 2: How would you change this algorithm to account for the possibility that the key is not found? What value(s) should be assigned to I?*

Ex. 3: Write a procedure that returns the value of the Nth item of a given file.

Since files are purely sequential data structures, only linear searches similar to this are applicable.

Often it is necessary to modify every item in a file. Since it is impossible to assign new values to the elements of an input file, or to do input and output at the same time with one file, we need a temporary file to hold the modified information until it can be copied back to the original file:

```
PROCEDURE  MODIFY( VAR F : ITEMFILE );
   (* Update all the items in the given file F.                *)

   VAR
      TEMP : ITEMFILE;   (* To hold the altered contents.   *)

   BEGIN (* MODIFY *)

   (* Prepare the file for I/O, then do the updating:          *)
   RESET( F );
   REWRITE( TEMP );
   WHILE NOT EOF(F) DO
      BEGIN (* each item *)

      (* Copy the entire item to the temporary file:           *)
      TEMP@ := F@;

      (* Update the various fields, as necessary:              *)
      (* (Assume that we just want to add 1 to INFO.)          *)
      TEMP@.INFO := TEMP@.INFO + 1;

      (* Move the pointers to the next value and location:*)
      PUT( TEMP );
      GET( F )
      END; (* each item *)

   (* Copy the changed file back to the original file:         *)
   COPYFILE( TEMP, F )

   END; (* MODIFY *)
```

Ex. 4: How would you change this algorithm so that only values of INFO (assuming integers, again) less than the value of CUTOFF are incremented?

Many algorithms use a temporary file this way—that is, to hold what will become the new contents of a given file.

†*Ex. 5: Write a SPLIT procedure that creates a file NEW out of all items in file OLD whose keys are in the range LOW..HIGH (assuming scalar keys), leaving OLD unchanged.*

†*Ex. 6: What changes must be made to SPLIT so that only the items with out-of-range keys are kept in file OLD?*

The items in a table are usually stored in order by their KEY values, as was shown in Part IV. The same is true for the items in a file. Assume, then, that the files are in ascending order by keys. We already know how to search for a given key and retrieve the associated information. We can also search an ordered file and insert a new item in its proper place:

```
PROCEDURE  INSERT( VAR F : ITEMFILE; I : ITEM );
   (* Insert a copy of I into the file F in ascending      *)
   (* order according to its KEY.                          *)

   VAR
      TEMP : ITEMFILE;   (* Holds the new file contents.   *)
      FOUND : BOOLEAN;   (* TRUE if a place for I is found.*)

   BEGIN (* INSERT *)

   (* Prepare for I/O, then search for the proper          *)
   (* location for item I:                                 *)
   RESET( F );
   REWRITE( TEMP );
   FOUND := FALSE;
   WHILE NOT EOF(F) AND NOT FOUND DO
      IF F@.KEY > I.KEY THEN
         (* The new item will be put at this point:        *)
         FOUND := TRUE
      ELSE
         BEGIN (* copy item *)
         TEMP@ := F@;
         PUT( TEMP );
         GET( F )
         END; (* copy item *)

   (* Insert the new item:                                 *)
   TEMP@ := I;
   PUT( TEMP );

   (* Copy the rest of F (if any) to TEMP, to make TEMP    *)
   (* complete:                                            *)
   COPYREST( F, TEMP );

   (* Copy the new contents back into F:                   *)
   COPYFILE( TEMP, F )

   END; (* INSERT *)
```

Ex. 7: Write a procedure to catenate (append) one file onto another.

†*Ex. 8: Write a procedure to delete the item with a given KEY value.*

Ex. 9: Write a procedure to delete all items whose keys are in the range LOW..HIGH.

The algorithms in this section are simple in concept (though not, perhaps, in implementation) because they are the basic operations performed on any homogeneous linear data structure. The next section presents some algorithms that are more often used with files than with other kinds of linear structures.

†*Ex. 10: The purpose of this exercise is to show how a large file can be constructed to allow some random access. Given the current declarations of ITEM, assume these also:*

```
CONST
   MAXITEMS = 50;

TYPE
   ITEMLIST =
      RECORD
      COUNT : 0..MAXITEMS;
      LIST : ARRAY[1..MAXITEMS] OF ITEM
      END;
   LISTFILE = FILE OF ITEMLIST;

VAR
   LF : LISTFILE;
   ILIST : ITEMLIST;
   I : ITEM;
```

LF is a file whose components are whole lists of items. Thus, 50 items are input (or output) at once, and each group can be sorted or can be searched with a fast binary search.

(a) Draw a picture of ILIST and of LF.
†*(b) Show how to initialize LF from data given as standard input.*
(c) Assume that the keys are steadily increasing throughout the file, and show how to search for the value of I in LF.
(d) Assuming ascending keys again, show how to insert the value of I into LF.
(e) Assume that each ITEMLIST in LF is initially half full; write an algorithm to insert a new value in ascending order (even when an ITEMLIST may have become full).

MERGES AND UPDATES

Probably the most important file algorithms (besides those for creation and printing) are the various kinds of merges and updates. This is because files are used as permanent databases to which new groups of data are added (merged) from time to time, or to which groups of "transactions" are applied (the file is "updated").

We know how to insert a single item into an ordered file. A *file merge* inserts an entire file full of items into another file. The original and resulting files are all ordered. In most cases the file containing the new information is merged into a *master* (permanent database) file, and the resulting file becomes the *new master:*

```
PROCEDURE  MERGE( VAR MASTER: ITEMFILE; VAR NEW: ITEMFILE );
   (* Merge the contents of the NEW file into MASTER to    *)
   (* create a new MASTER.                                  *)

   VAR
      TEMP : ITEMFILE;   (* Holds the merged contents.      *)

   BEGIN (* MERGE *)

   (* Prepare the files for I/O:                            *)
   RESET( NEW );
   RESET( MASTER );
   REWRITE( TEMP );

   (* Step through both input sequences, always putting     *)
   (* the item with the lesser KEY into TEMP:               *)
   (* (TEMP will therefore be ordered.)                     *)
   WHILE NOT EOF(MASTER) AND NOT EOF(NEW) DO
      BEGIN (* each pair *)
      IF MASTER@.KEY < NEW@.KEY THEN
         BEGIN (* take from MASTER *)
         TEMP@ := MASTER@;
         PUT( TEMP );
         GET( MASTER )
         END (* take from MASTER *)
      ELSE
         BEGIN (* take from NEW *)
         TEMP@ := NEW@;
         PUT( TEMP );
         GET( NEW )
         END (* take from new *)
      END; (* each pair *)

   (* Continued on the next page...                         *)
```

```
(* Assuming that the NEW file is at EOF, copy the rest *)
(* of MASTER over to TEMP:                              *)
COPYREST( MASTER, TEMP );

(* It might have been MASTER that was at EOF:           *)
COPYREST( NEW, TEMP );

(* Now copy the completed file back to the MASTER file:*)
COPYFILE( TEMP, MASTER )

END; (* MERGE *)
```

†*Ex. 11: The MERGE procedure does not check to see if both files have any items with identical KEY values, so the MASTER file may now contain duplicate items. What changes can be made to MERGE to avoid this outcome (if necessary)?*

The other major file algorithm, the *file update,* combines insertions, deletions, and changes to items in a single algorithm. Typically, the action to be performed with a given item is called a *transaction.* There is always a *transaction file,* ordered by keys, that contains items to be inserted into the MASTER file (as in a merge) and items that are used to indicate changes to (or deletions of) items in the MASTER. The transaction file is assumed to have been created by another algorithm prior to the intended update.

This is how the transactions might be applied to the MASTER file, yielding an updated MASTER:

```
PROCEDURE  UPDATE( VAR MASTER : ITEMFILE;
                   VAR TRANS : ITEMFILE );
   (* Apply the given updating transactions from TRANS to *)
   (* MASTER.  The actions to be taken are:               *)
   (*    1. All items in TRANS that are not yet in MASTER *)
   (*       should be inserted into MASTER.               *)
   (*    2. A key in TRANS matching one in MASTER means:  *)
   (*       a) delete the item in MASTER if the INFO field *)
   (*          has the special (previously agreed upon)   *)
   (*          value of DROPIT,                           *)
   (*       b) otherwise just change the INFO field in the *)
   (*          MASTER record to the value of the one in   *)
   (*          this TRANS record.                         *)

   VAR
      TEMP : ITEMFILE;  (* Holds updated file contents.   *)

   BEGIN (* UPDATE *)

   (* Ready the files:                                    *)
   RESET( MASTER );
   RESET( TRANS );
   REWRITE( TEMP );

   (* Continued on the next page...                       *)
```

```
(* Step through both files, finding keys that match,    *)
(* or finding where a key is to be inserted (merged):   *)
WHILE NOT EOF(MASTER) AND NOT EOF(TRANS) DO
   IF TRANS@.KEY < MASTER@.KEY THEN
      BEGIN (* insert *)
      (* Insert the new value, then get a transaction: *)
      TEMP@ := TRANS@;
      PUT( TEMP );
      GET( TRANS )
      END (* insert *)
   ELSE IF TRANS@.KEY = MASTER@.KEY THEN
      BEGIN (* change or delete *)
      (* Apply the change as if it were not a delete:  *)
      TEMP@ := MASTER@;
      TEMP@.INFO := TRANS@.INFO;
      GET( MASTER );
      GET( TRANS );
      (* Keep this transaction if it is not a delete:  *)
      IF TEMP@.INFO <> DROPIT THEN
         PUT( TEMP )
      END (* change or delete *)
   ELSE (* TRANS@.KEY > MASTER@.KEY *)
      BEGIN (* MASTER unchanged *)
      (* The transaction's key is beyond this point:   *)
      TEMP@ := MASTER@;
      PUT( TEMP );
      GET( MASTER )
      END; (* MASTER unchanged *)

(* Copy the rest of the MASTER file, if any:            *)
COPYREST( MASTER, TEMP );

(* If (instead) any transactions are left, they should *)
(* be insertions:                                       *)
COPYREST( TRANS, TEMP );

(* Copy the updated file back into MASTER:              *)
COPYFILE( TEMP, MASTER )

END; (* UPDATE *)
```

†*Ex. 12: Why should leftover transactions be insertions only? What changes would you make in case they were not?*

Ex. 13: Show how the transaction file could have been created from standard input, where the data for each intended transaction is preceded by 'A', 'C', or 'D', meaning add, change, or delete, respectively, and DROPIT has the value -MAXINT.

Note that UPDATE and MERGE both used file pointers and had similar structure. Each also applied one file to another, changing the latter. Either algorithm can be generalized to combine any two files to obtain a third. The following file merge does so. More importantly, it shows that a merge can be done using extra input variables instead of the file pointers. It does not use the COPYREST procedure:

```
PROCEDURE  MERGE( VAR FIRST : ITEMFILE;
                  VAR SECOND : ITEMFILE;
                  VAR NEW : ITEMFILE );
   (* Merge the ordered files FIRST and SECOND into NEW     *)
   (* (in ascending order).  ***The keys are assumed to     *)
   (* be integers less than MAXINT.                         *)

   VAR
      FIRSTITEM : ITEM; (* An element of FIRST.             *)
      SECONDITEM : ITEM;(* An element of SECOND.            *)

   PROCEDURE  GETITEM( VAR F : ITEMFILE; VAR I : ITEM );
      (* Read an item from F into I, if possible, setting *)
      (* I's key to MAXINT if F is empty.                 *)

      BEGIN (* GETITEM *)
      IF NOT EOF(F) THEN
         READ( F, I )
      ELSE
         I.KEY := MAXINT
      END; (* GETITEM *)

   BEGIN (* MERGE *)

   RESET( FIRST );
   RESET( SECOND );
   REWRITE( NEW );

   (* Compare key values from the two input files, always *)
   (* copying the item with the smaller key to the output:*)
   WHILE (FIRSTITEM <> MAXINT) OR (SECONDITEM <> MAXINT) DO
      IF FIRSTITEM.KEY < SECONDITEM.KEY THEN
         BEGIN (* take from FIRST *)
         WRITE( NEW, FIRSTITEM );
         GETITEM( FIRST, FIRSTITEM )
         END (* take from FIRST *)
      ELSE
         BEGIN (* take from SECOND *)
         WRITE( NEW, SECONDITEM );
         GETITEM( SECOND, SECONDITEM )
         END (* take from SECOND *)

   END; (* MERGE *)
```

Ex. 14: Write a file update based upon this merging strategy.

SORTING

It may at first be surprising to learn that the contents of a file can be sorted without first being stored in an array or table, but it *is* possible. This section presents one of the many algorithms for doing just that. (Others can be found in the algorithm texts listed in the Bibliography.) This one is based on the fact that any file contains successive sequences of values (called *runs)* that are in ascending order already. (There might be only one value in a run!) If each pair of these runs can be "held" side-by-side, they can be merged into a single, longer run. Pairs of the longer runs can then be merged, and so on; a sorted file is one whose values have all been recombined into one run:

```
PROCEDURE  SORT( VAR F : ITEMFILE );
   (* Sort the file F into ascending order, using the     *)
   (* "natural merge sort" algorithm.                     *)

   VAR
      RUNS : 0..MAXINT; (* Number of sequences of         *)
                        (* ascending order values.        *)
      TEMP1 : ITEMFILE; (* These hold runs of items that  *)
      TEMP2 : ITEMFILE; (* are from F, to be compared and *)
                        (* merged back into F.            *)
      ENDRUN : BOOLEAN; (* TRUE only at the end of a run, *)
                        (* while reading any file.        *)

   PROCEDURE  COPYITEM( VAR SOURCE : ITEMFILE;
                        VAR DEST : ITEMFILE );
      (* Copy a single value from SOURCE to DEST, setting *)
      (* the global ENDRUN to TRUE if this value is the   *)
      (* last in an ascending sequence (run).             *)

      VAR
         I : ITEM;      (* The value being copied.        *)

      BEGIN (* COPYITEM *)
      READ( SOURCE, I );
      WRITE( DEST, I );
      (* Check for end of run:                            *)
      IF EOF(SOURCE) THEN
         ENDRUN := TRUE
      ELSE
         ENDRUN := (I.KEY > SOURCE@.KEY)
      END; (* COPYITEM *)

      (* Continued on the next page...                    *)
```

```
PROCEDURE  COPYRUN( VAR SOURCE : ITEMFILE;
                    VAR DEST : ITEMFILE );
   (* Copy a run (ascending sequence) of values from   *)
   (* SOURCE  to DEST.                                  *)

   BEGIN (* COPYRUN *)
   REPEAT
      COPYITEM( SOURCE, DEST )
   UNTIL ENDRUN       (* ENDRUN is set by COPYITEM.     *)
   END; (* COPYRUN *)

PROCEDURE  SPLITUP( VAR F : ITEMFILE;
                    VAR TEMP1 : ITEMFILE;
                    VAR TEMP2 : ITEMFILE );
   (* Copy the successive runs from F into the files    *)
   (* TEMP1 and TEMP2, alternating whole runs between   *)
   (* the two files.                                    *)

   BEGIN (* SPLITUP *)
   RESET( F );
   REWRITE( TEMP1 );
   REWRITE( TEMP2 );
   REPEAT
      COPYRUN( F, TEMP1 );
      IF NOT EOF(F) THEN
         COPYRUN( F, TEMP2 )
   UNTIL EOF(F)
   END; (* SPLITUP *)

PROCEDURE  MERGERUNS( VAR TEMP1 : ITEMFILE;
                      VAR TEMP2 : ITEMFILE;
                      VAR F : ITEMFILE );
   (* Merge the next pair of runs from TEMP1 and TEMP2 *)
   (* into a single (and longer) run back into F.      *)

   BEGIN (* MERGERUNS *)
   REPEAT
      IF TEMP1@.KEY < TEMP2@.KEY THEN
         BEGIN (* take from TEMP1 *)
         COPYITEM( TEMP1, F );
         (* The run in TEMP1 could be the shorter one: *)
         IF ENDRUN THEN
            COPYRUN( TEMP2, F )
         END (* take from TEMP1 *)
      ELSE
         BEGIN (* take from TEMP2 *)
         COPYITEM( TEMP2, F );
         (* The run in TEMP2 may be the shorter:       *)
         IF ENDRUN THEN
            COPYRUN( TEMP1, F )
         END (* take from TEMP2 *)
   UNTIL ENDRUN
   END; (* MERGERUNS *)
```

```
PROCEDURE  RECOMBINE( VAR TEMP1 : ITEMFILE;
                      VAR TEMP2 : ITEMFILE;
                      VAR F : ITEMFILE );
   (* Merge all pairs of runs from TEMP1 and TEMP2,     *)
   (* combining them into successive (and longer) runs *)
   (* in F.  Count the number of runs in F.             *)

   BEGIN (* RECOMBINE *)

   RESET( TEMP1 );
   RESET( TEMP2 );
   REWRITE( F );

   (* Merge and count the pairs of runs:                *)
   WHILE NOT EOF(TEMP1) AND NOT EOF(TEMP2) DO
      BEGIN (* each pair *)
      MERGERUNS( TEMP1, TEMP2, F );
      RUNS := RUNS + 1
      END; (* each pair *)

   (* There may be some runs left in one of the files: *)
   WHILE NOT EOF(TEMP1) DO
      BEGIN
      COPYRUN( TEMP1, F );
      RUNS := RUNS + 1
      END;
   WHILE NOT EOF(TEMP2) DO
      BEGIN
      COPYRUN( TEMP2, F );
      RUNS := RUNS + 1
      END
   END; (* RECOMBINE *)


BEGIN (* SORT *)

(* Split the file into its successive runs, then      *)
(* recombine pairs of runs into longer runs, until    *)
(* there is only one run:                             *)
REPEAT
   SPLITUP( F, TEMP1, TEMP2 );
   RUNS := 0;
   RECOMBINE( TEMP1, TEMP2, F )
UNTIL RUNS = 1

END; (* SORT *)
```

Ex. 15: Why are the runs held in files instead of in arrays?

†*Ex. 16: Given an ordered file OLD and an unordered file NEW, write a procedure to merge NEW into OLD in order without first sorting NEW.*

OTHER OPERATIONS ON FILES

Many other operations are performed upon files. This section presents a few more algorithms so that you can see how wide-ranging these operations are.

First, however, it is well to note the use of a programming strategy frequently called *spooling*. Spooling is simply using a file to hold a list of values temporarily. This is usually how arrays are used. It is also what has already been shown in many of the file algorithms, but the emphasis was always upon holding the contents of other *files* temporarily. The application here is more like arrays.

Assume that we have these declarations:

```
VAR
    A : ARRAY[1..MAX] OF INTEGER;
    F : FILE OF INTEGER;
```

The implication is that both A and F are just different kinds of linear data structures in the program. If the program generates a list of integers that it will need later, for example, we can choose the appropriate structure depending upon how big the list is and how it is to be accessed. Both structures are linear lists of the same type of elements, and both can be manipulated with many similar algorithms, but the array permits immediate access to any element, whereas the file's advantage is that it can hold any number of elements. Programmers should feel comfortable enough with files to use them (when they are appropriate) just like other program variables.

Ex. 17: (a) Write a program that takes as input an unknown number of integers. Have it print the difference between each integer and the overall average. (Use a spooling file.) (b) Write a program that reads an unknown number of integers and prints all positive values before it prints the negative values. These are examples of problems that can not be solved without some kind of unbounded data structure. (Note that the standard files INPUT and OUTPUT are unbounded data structures.)

To continue with other operations on files, note that a file might contain items with duplicate keys. This could happen if it had been created from unordered data and then sorted, or as the result of a merge, or for some other reason. In any case, there is an algorithm to remove the duplicates, if necessary (it often is not):

```
PROCEDURE  COMPRESS( VAR F : ITEMFILE );
   (* Remove items with duplicate keys from the file F.   *)

   VAR
      TEMP : ITEMFILE;   (* To save the compressed file.   *)
      HOLD : ITEM;       (* For comparison with next item. *)
      ANY : BOOLEAN;     (* TRUE if the file is not empty. *)

   BEGIN (* COMPRESS *)

   (* Prepare the files for I/O:                           *)
   RESET( F );
   REWRITE( TEMP );

   (* Get and hold the first item so that it can be        *)
   (* compared with the next (perhaps equal) one:          *)
   ANY := NOT EOF(F);
   IF ANY THEN
      READ( F, HOLD );

   (* Step through the items, writing them only when the   *)
   (* key changes:                                         *)
   WHILE NOT EOF(F) DO
      (* Compare the next key to the one held back:        *)
      IF F@.KEY <> HOLD.KEY THEN
         BEGIN (* key change *)
         (* Write the one held back, then hold the next:   *)
         WRITE( TEMP, HOLD );
         READ( F, HOLD )
         END (* key change *)
      ELSE
         (* Same key, so just skip it:                     *)
         GET( F );

   (* If there were any at all, output the item held back:*)
   IF ANY THEN
      WRITE( TEMP, HOLD );

   (* Copy the compressed file back to the original file: *)
   COPYFILE( TEMP, F )

   END; (* COMPRESS *)
```

†*Ex. 18: Assuming that INFOTYPE is integer, show the changes to COMPRESS that will make it sum the INFO fields of each set of items having the same key, saving that sum in the INFO field of the one item retained in the file.*

Like arrays, files can be compared item by item to see if they are the same:

```
FUNCTION  COMPARE( VAR ONE : ITEMFILE;
                   VAR OTHER : ITEMFILE ) : BOOLEAN;
   (* Return TRUE only if there is a 1-to-1                *)
   (* correspondence between the keys in ONE and OTHER.    *)

   VAR
      MATCH : BOOLEAN;   (* TRUE while the items match.    *)

   BEGIN (* COMPARE *)

   (* Ready the files for I/O:                             *)
   RESET( ONE );
   RESET( OTHER );

   (* See if the existing items in the files match:        *)
   MATCH := TRUE;
   WHILE NOT ( EOF(ONE) OR EOF(OTHER) ) AND MATCH DO
      IF ONE@.KEY <> OTHER@.KEY THEN
         MATCH := FALSE
      ELSE
         BEGIN (* step to next *)
         GET( ONE );
         GET( OTHER )
         END; (* step to next *)

   (* Both files must also now be at EOF, or else there    *)
   (* is not a 1-to-1 correspondence between their keys:   *)
   COMPARE := MATCH AND EOF(ONE) AND EOF(OTHER)

   END; (* COMPARE *)
```

Ex. 19: Think about an algorithm that compares two versions of the "same" file to find out where they differ. Either file could have extra records that the other lacks. (This is a very difficult problem.)

Finally, it is even possible to use a file to implement a stack or a queue, because they are all linear data structures. The only advantage would be that the file may represent a permanent data file, so that the stack or queue would persist between executions of the program. The only reason to look at such an algorithm here is to see that such things are possible because of the close relationship of all linear data structures.

Assume, for example, that we have declared the type STACK as FILE OF INTEGER, and that the stack top is the first integer in the file. A stack POP operation can be done this way:

```
PROCEDURE  POP( VAR STK : STACK; VAR I : INTEGER );
   (* Remove the top value of the given stack, returning  *)
   (* it as the value of I.                                *)

   VAR
      TEMP : STACK;      (* Holds all but the top value.   *)

   BEGIN (* POP *)

   (* Get the value on top:                                *)
   RESET( STK );
   READ( STK, I );

   (* Save the rest:                                       *)
   REWRITE( TEMP );
   COPYREST( STK, TEMP );
   COPYFILE( TEMP, STK )

   END; (* POP *)
```

Ex. 20: Write procedures or functions to push a new value onto the stack, to check the stack for empty, and to compare the top value of the stack to a given value.

PROGRAMS FOR READING

```
 1  PROGRAM  SAVECHECKS( INPUT, OUTPUT, ALLCHKS );
 2
 3  (* PURPOSE:                                                      *)
 4  (*    Build a database of personal checks.                       *)
 5  (*                                                               *)
 6  (* PROGRAMMER:   David V. Moffat                                 *)
 7  (*                                                               *)
 8  (* INPUT:                                                        *)
 9  (*    Check information from a check register, on file           *)
10  (*    INPUT, like this:                                          *)
11  (*       col. 1-3:    check number (integer)                     *)
12  (*       col. 4:      blank                                      *)
13  (*       col. 5-11:   amount spent (real)                        *)
14  (*       col. 12:     blank                                      *)
15  (*       col. 13-18: date, given as mmddyy (char)                *)
16  (*       col. 19:     blank                                      *)
17  (*       col. 20-eoln: category of expenditure (10 chars)        *)
18  (*                                                               *)
19  (* OUTPUT:                                                       *)
20  (*    File ALLCHKS:  Records containing the information,         *)
21  (*       one per check.                                          *)
22  (*    File OUTPUT:  An echo of the data.                         *)
23  (*                                                               *)
24  (* ASSUMPTIONS & LIMITATIONS:                                    *)
25  (*    The data are assumed to be correct and complete.           *)
26  (*                                                               *)
27  (* ALGORITHM:                                                    *)
28  (*    Initialize file and print titles.                          *)
29  (*    Get and echo the data until EOF:                           *)
30  (*       Read and echo check information.                        *)
31  (*       Write a record to ALLCHKS.                              *)
32  (*    Summarize.                                                 *)
33
```

```
34  TYPE
35     MMDDYY = PACKED ARRAY[1..6] OF CHAR;
36                             (* 6 digits of a date.              *)
37     CATEGORY = PACKED ARRAY[1..10] OF CHAR;
38                             (* What check was written for.      *)
39     CHECKTYPE =             (* One check register entry:        *)
40        RECORD
41        NUMBER : 100..999;(* Always a three digit number.        *)
42        AMOUNT : REAL;       (* Dollars and cents.               *)
43        DATE : MMDDYY;       (* Date check was written.          *)
44        EXPENSE : CATEGORY   (* What it was written for.         *)
45        END;
46
47  VAR
48     CHECK : CHECKTYPE;
49                             (* A single check (from INPUT).     *)
50     ALLCHKS : FILE OF CHECKTYPE;
51                             (* The new database of checks.      *)
52
53  PROCEDURE  GETANDECHO( VAR CHECK : CHECKTYPE );
54     (* Reads check information from INPUT and puts it          *)
55     (* into a record for ALLCHKS.  Echoes the information.     *)
56
57     CONST                   (* To blank out categories:         *)
58        BLANKCAT = '             ';
59
60     VAR
61        CH : CHAR;           (* One character of input.          *)
62        I : INTEGER;         (* FOR loop index.                  *)
63
64     BEGIN (* GETANDECHO *)
65     WITH CHECK DO
66        BEGIN
67        READ( INPUT, NUMBER, AMOUNT, CH (*blank*) );
68        FOR I:=1 TO 6 DO
69           BEGIN
70           READ( INPUT, CH );
71           DATE[I] := CH
72           END;
73        READ( INPUT, CH (*blank*) );
74        EXPENSE := BLANKCAT;
75        I := 0;
76        WHILE (I < 10) AND NOT EOLN(INPUT) DO
77           BEGIN (* each char *)
78           READ( INPUT, CH );
79           I := I + 1;
80           EXPENSE[I] := CH
81           END;
82        READLN( INPUT );
83        WRITELN( OUTPUT, NUMBER:3, AMOUNT:9:2, DATE:8,
84                 EXPENSE:12 )
85        END
86     END; (* GETANDECHO *)
```

```
 87
 88
 89   BEGIN (* SAVECHECKS *)
 90
 91   (* Initialize file for the database, print OUTPUT titles: *)
 92   REWRITE( ALLCHKS );
 93   PAGE( OUTPUT );
 94   WRITELN( OUTPUT, 'Echo of Check Register Data:' );
 95   WRITELN( OUTPUT );
 96   WRITELN( OUTPUT, '#':3, 'Amount':9, 'mmddyy':8,
 97            'Category':12 );
 98
 99   (* Get and echo the data until EOF on INPUT:             *)
100   WHILE NOT EOF(INPUT) DO
101      BEGIN (* each check *)
102      GETANDECHO( CHECK );
103      WRITE( ALLCHKS, CHECK )
104      END; (* each check *)
105
106   (* Summarize:                                            *)
107   WRITELN( OUTPUT );
108   WRITELN( OUTPUT, 'End of data.':21 );
109   PAGE( OUTPUT )
110   END. (* SAVECHECKS *)
```

```
Echo of Check Register Data:

  #    Amount   mmddyy    Category
469     10.66   070282  FOODOUT
470     20.80   070382  GROCERIES
471     26.85   070382  GROCERIES
472    200.00   070582  RENT
473     92.81   070582  TELEPHONE
474     15.30   070582  ELECTRIC
475     41.02   070582  PIANO
476     88.33   070582  LOAN
477    122.00   070582  AUTOINSURA
478      7.96   071682  FOODOUT
479     29.21   071782  GROCERIES
480     18.00   071882  MOTORCLUB
481     16.64   071882  ELECTRIC
482     18.04   071882  FOODOUT
483    714.54   072282  LEGALFEES
484      6.91   072382  ASPIRINS
485     17.98   072382  GROCERIES
486     14.77   072482  HOUSEHOLD
487      5.56   072482  HARDWARE
488      9.15   072482  FOODOUT
489     14.32   072482  DRUGSTORE
490      6.91   072582  FOODOUT

         End of data.
```

```
 1   PROGRAM  SELECT( INPUT, OUTPUT, ALLCHKS, CHKOUT );
 2
 3   (* PURPOSE:                                                *)
 4   (*    Select checks with a given expense category from a   *)
 5   (*    database, and put them onto another file.            *)
 6   (*                                                         *)
 7   (* PROGRAMMER:  David V. Moffat                            *)
 8   (*                                                         *)
 9   (* INPUT:                                                  *)
10   (*    File ALLCHKS: Database of checks (created by the     *)
11   (*       program SAVECHECKS).                              *)
12   (*    File INPUT: An expense category of checks to be      *)
13   (*       selected from the database.  Format:              *)
14   (*          col.1-eoln: category (10 characters are kept) *)
15   (*    File CHKOUT: The checks selected from ALLCHKS.       *)
16   (*    File OUTPUT: Echo of the category and File CHKOUT.  *)
17   (*                                                         *)
18   (* ASSUMPTIONS & LIMITATIONS:                              *)
19   (*    The selection category is given correctly.           *)
20   (*                                                         *)
21   (* ALGORITHM:                                              *)
22   (*    Initialize.                                          *)
23   (*    Get and echo the category from INPUT.                *)
24   (*    Select all records in ALLCHKS with the given         *)
25   (*       category and write them out to CHKOUT.            *)
26   (*    Print the contents of CHKOUT, if any.                *)
27
28   TYPE
29      MMDDYY = PACKED ARRAY[1..6] OF CHAR;
30                              (* 6 digits of a date.          *)
31      CATEGORY = PACKED ARRAY[1..10] OF CHAR;
32                              (* What the check was written for.*)
33      CHECKTYPE =             (* One check register entry:    *)
34         RECORD
35         NUMBER : 100..999;(* Always a three digit number.   *)
36         AMOUNT : REAL;     (* Dollars and cents.             *)
37         DATE : MMDDYY;     (* Date check was written.        *)
38         EXPENSE : CATEGORY   (* What it was written for.     *)
39         END;
40
41   VAR
42      ALLCHKS : FILE OF CHECKTYPE;
43                              (* Database of check information. *)
44      CHKOUT : FILE OF CHECKTYPE;
45                              (* All checks with given category.*)
46      GIVENCAT : CATEGORY; (* Category to select from ALLCHKS*)
47      I : INTEGER;            (* Index into GIVENCAT.          *)
48      CH : CHAR;              (* One character of input.       *)
49
```

```
 50   BEGIN (* SELECT *)
 51
 52   PAGE( OUTPUT );
 53
 54   (* Get and echo the category from INPUT:                        *)
 55   GIVENCAT := '          ';
 56   I := 0;
 57   WHILE (I < 10) AND NOT EOLN(INPUT) DO
 58      BEGIN (* each char *)
 59      READ( INPUT, CH );
 60      I := I + 1;
 61      GIVENCAT[I] := CH
 62      END; (* each char *)
 63   WRITELN( OUTPUT, 'The expense category to be selected is ',
 64            '"', GIVENCAT, '"' );
 65
 66   (* Initialize files for I/O:                                    *)
 67   RESET( ALLCHKS );
 68   REWRITE( CHKOUT );
 69
 70   (* Select all records in ALLCHKS with the given category, *)
 71   (* writing them to CHKOUT:                                      *)
 72   WHILE NOT EOF(ALLCHKS) DO
 73      BEGIN (* each check *)
 74      IF ALLCHKS@.EXPENSE = GIVENCAT THEN
 75         BEGIN (* write it *)
 76         CHKOUT@ := ALLCHKS@;
 77         PUT( CHKOUT )
 78         END; (* write it *)
 79      GET( ALLCHKS )
 80      END; (* each check *)
 81
 82   (* Print the contents of CHKOUT, if any:                        *)
 83   WRITELN( OUTPUT, 'The checks selected are:' );
 84   WRITELN( OUTPUT );
 85   RESET( CHKOUT );
 86   IF EOF(CHKOUT) THEN
 87      WRITELN( OUTPUT, '(There were none selected.)' )
 88   ELSE
 89      BEGIN (* print it *)
 90      WRITELN( OUTPUT, '#':3, 'Amount':9, 'mmddyy':8,
 91               'Category':12 );
 92      REPEAT
 93         WITH CHKOUT@ DO
 94            WRITELN( OUTPUT, NUMBER:3, AMOUNT:9:2, DATE:8,
 95                     EXPENSE:12 );
 96         GET( CHKOUT )
 97      UNTIL EOF(CHKOUT)
 98      END; (* print it *)
 99   WRITELN( OUTPUT );
100   WRITELN( OUTPUT, '***End of processing.' );
101   PAGE( OUTPUT )
102   END. (* SELECT *)
```

```
The expense category to be selected is "FOODOUT   "
The checks selected are:

  #    Amount   mmddyy     Category
469     10.66   070282   FOODOUT
478      7.96   071682   FOODOUT
482     18.04   071882   FOODOUT
488      9.15   072482   FOODOUT
490      6.91   072582   FOODOUT

***End of processing.
```

PROGRAM EXERCISES

1. Show all changes that must be made to program SELECT so that it will input several selection categories, then build CHKOUT from all checks written for any of the given categories.

2. What changes must be made to procedure GETANDECHO of program SAVECHECKS to allow the input of checks without a category?

Part VII: Pointer Algorithms

INTRODUCTION

Pointers are the most versatile of all the types. Their great utility is due to two properties: they can be used to allocate on demand variables of any type, and they can be used to build data structures that represent an unlimited variety of real or conceptual objects.

There are two situations in which allocation on demand is advantageous: when a large data structure (such as an array) may be needed only briefly during the execution of a program, and when the structure may or may not be used from one run to the next, depending upon the particular data. To keep a large data structure around at all times can make the space required by the program unnecessarily large. As an alternative, the data structure can be declared as the object of a pointer type:

```
VAR
    RA : @ARRAY[1..1000] OF REAL;
```

(The "@" is the standard alternative to the caret. "@" is easier to see and more mnemonic.)

We can choose when to create and use this array—for example, depending upon the result of a test:

```
IF NEEDARRAY THEN
    BEGIN (* allocate and use *)
    NEW( RA );
    ...;
    RA@[I] := ...;
    ...;
    DISPOSE( RA )
    END (* allocate and use *)
```

This strategy also has an advantage over the use of a file to hold the same information temporarily: access to the elements of the array is faster and can be done at random. Of course, the size of the array must be written into the program, whereas a file is essentially unbounded.

The more interesting property of the pointer type, its use in *building* data structures, is the subject of the next several sections. Most of the data structures require these declarations:

```
TYPE
   INFOTYPE = ...;          (* Information to be stored.         *)
   NODEPTR = @NODE;         (* The basic pointer to elements.  *)
   NODE =                   (* One element of a structure:     *)
      RECORD
      INFO : INFOTYPE;      (* This element's information.      *)
      NEXT : NODEPTR        (* Pointer to another element.      *)
      END;
```

Because each element or "node" can point or "link" to another node, as in the following diagram, the data structures that are represented with these elements are collectively called *linked lists*.

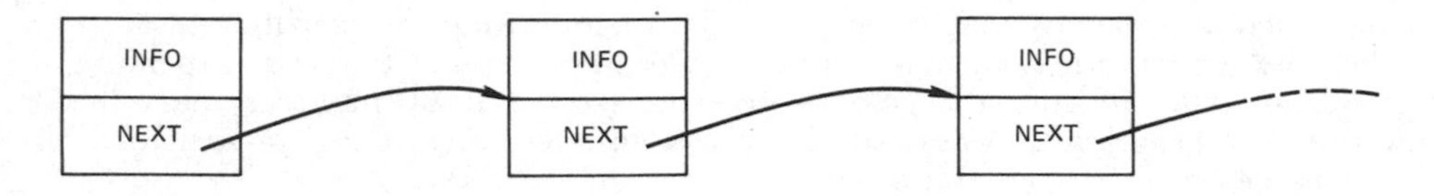

STACKS

A stack is a linear data structure that is accessed from only one end. Stacks can be implemented using arrays, as in Part III, or even with files. They are often best implemented as linked lists so that they will be unbounded and efficient.

Assume that we have these declarations in addition to those given in the Introduction:

```
TYPE
   STACK =
      RECORD
      TOP : NODEPTR        (* Pointer to the STACK's "top".   *)
      END;

VAR
   STK : STACK;
   ITEM : INFOTYPE;
```

(Because the STACK type is defined as a record, every variable of this type has a TOP.)

The stack STK can hold as much information as any number of ITEMs. This diagram shows an empty stack and a stack containing three nodes of information:

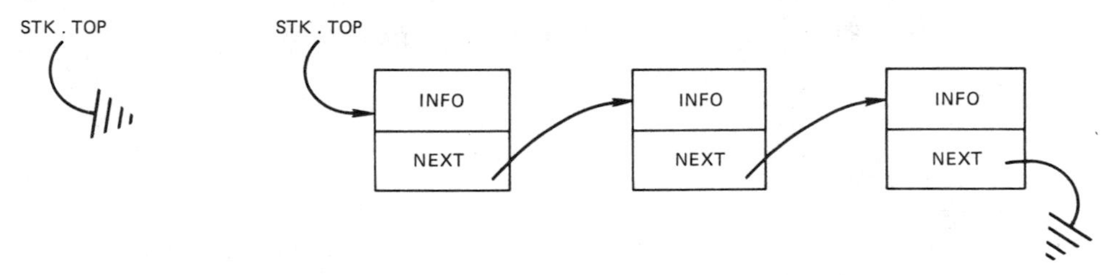

The peculiar-looking arrows in the diagram represent pointers whose values are NIL—they point to nothing but *are* initialized.

A stack can be initialized to empty (having no elements) this way:

```
PROCEDURE  STACKINIT( VAR STK : STACK );
   (* Initialize a given stack to empty.                      *)

   BEGIN (* STACKINIT *)

   STK.TOP := NIL

   END; (* STACKINIT *)
```

Information is placed into a stack or removed from it only at its top. A new value is "pushed" onto the stack like this:

```
PROCEDURE  PUSH( ITEM : INFOTYPE; VAR STK : STACK );
   (* Push the value of ITEM onto the top of STK.             *)

   VAR
      P : NODEPTR;        (* A temporary pointer.             *)

   BEGIN (* PUSH *)

   (* Create a new node to hold the value of ITEM:            *)
   NEW( P );
   P@.INFO := ITEM;
   (* The new node becomes the new top of the stack:          *)
   P@.NEXT := STK.TOP;
   STK.TOP := P

   END; (* PUSH *)
```

Conversely, the top value is removed ("popped") from the top of the stack this way:

```
PROCEDURE  POP( VAR STK : STACK; VAR ITEM : INFOTYPE );
   (* Pop the top element of the given stack, and return      *)
   (* it as the value of ITEM:                                *)

   VAR
      P : NODEPTR;        (* A temporary pointer.             *)

   BEGIN (* POP *)

   (* Copy the value of the top node into ITEM:               *)
   P := STK.TOP;
   ITEM := P@.INFO;
   (* Get rid of the current top node, exposing the next: *)
   STK.TOP := P@.NEXT;
   DISPOSE( P )

   END; (* POP *)
```

Ex. 1: What rule of style dictates that we not use a function for POP?

Of course, a value can be popped from a stack only if there are nodes remaining in it. We can see if a stack is empty with this function:

```
FUNCTION  STACKEMPTY( STK : STACK ) : BOOLEAN;
   (* Return TRUE if the given stack is empty.                *)

   BEGIN (* STACKEMPTY *)

   STACKEMPTY := ( STK.TOP = NIL )

   END; (* STACKEMPTY *)
```

We often wish to inspect the value on the top of a stack without removing it. To see if the stack's top value is equal to some other value, we can use:

```
FUNCTION  EQUALTOP( ITEM : INFOTYPE; STK : STACK ): BOOLEAN;
   (* Return TRUE if the value of ITEM is equal to the    *)
   (* information in the top element of the given stack.  *)

   BEGIN (* EQUALTOP *)

   EQUALTOP := ( ITEM = STK.TOP@.INFO )

   END; (* EQUALTOP *)
```

†*Ex. 2: Rewrite POP and EQUALTOP so that each checks for an empty stack itself. What value should each return if the stack is empty?*

QUEUES

A queue is another data structure that can be implemented as a linked list. It is used to represent a waiting list or line. (In England, a line of commuters waiting for a bus is called a queue.) Elements are put into a queue at one end and removed from the other.

A queue can be defined this way:

```
TYPE
   QUEUE =
      RECORD
      FRONT : NODEPTR;    (* Pointer to the first in line.  *)
      REAR : NODEPTR      (* Pointer to the last in line.   *)
      END;

VAR
   WAITQ : QUEUE;
   ITEM : INFOTYPE;
```

The queue WAITQ can store the values of zero or more ITEMs. An empty queue and a queue of three values would look like this:

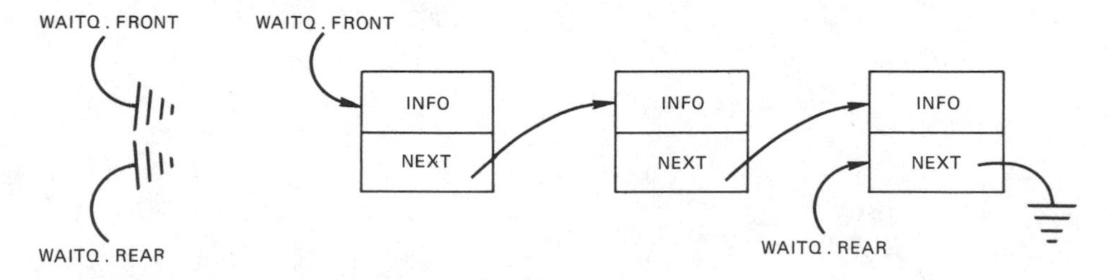

A queue can be initialized to empty using this procedure:

```
PROCEDURE  QINIT( VAR WAITQ : QUEUE );
   (* Initialize the given queue to empty.                  *)

   BEGIN (* QINIT *)

   WAITQ.FRONT := NIL;
   WAITQ.REAR := NIL

   END; (* QINIT *)
```

A queue differs from a stack in the way it is used. Nodes containing new information are always inserted ("enqueued") at the rear of the queue:

```
PROCEDURE  ENQUEUE( ITEM : INFOTYPE; VAR Q : QUEUE );
   (* Insert the value of the given ITEM at the rear of    *)
   (* the given queue.                                     *)

   VAR
      P : NODEPTR;        (* A temporary pointer.          *)

   BEGIN (* ENQUEUE *)

   (* Create a node for the new value:                     *)
   NEW( P );
   P@.INFO := ITEM;
   P@.NEXT := NIL;

   (* Link it into the rear end of the queue:              *)
   IF Q.FRONT = NIL THEN
      (* The queue is empty.  This will be the only node: *)
      Q.FRONT := P
   ELSE
      (* The current rear node must now point to this one:*)
      Q.REAR@.NEXT := P;

   (* In any case, this node is now at the rear:           *)
   Q.REAR := P

   END; (* ENQUEUE *)
```

Information is always taken out from the front end of the queue:

```
PROCEDURE  REMOVE( VAR Q : QUEUE; VAR ITEM : INFOTYPE );
   (* Remove the node from the front of the given queue,  *)
   (* returning its value as the value of ITEM.           *)

   VAR
      P : NODEPTR;        (* A temporary pointer.          *)

   BEGIN (* REMOVE *)

   (* Copy the value from the front node:                  *)
   P := Q.FRONT;
   ITEM := P@.INFO;

   (* Get rid of it, exposing the node next in line:       *)
   Q.FRONT := P@.NEXT;
   IF Q.FRONT = NIL THEN
      Q.REAR := NIL;

   DISPOSE( P )

   END; (* REMOVE *)
```

Once again, a value can be removed only if any remain in the queue, so we need a function to check for an empty queue:

```
FUNCTION  QEMPTY( Q : QUEUE ) : BOOLEAN;
   (* Returns TRUE if the given queue is empty.                *)

   BEGIN (* QEMPTY *)

   QEMPTY := ( Q.FRONT = NIL )

   END; (* QEMPTY *)
```

Finally, we often want to print the contents of a queue:

```
PROCEDURE  PRINTQUEUE( Q : QUEUE );
   (* Print all the values stored in the given queue,        *)
   (* from front to rear.                                    *)

   VAR
      P : NODEPTR;         (* For stepping through the list. *)

   BEGIN (* PRINTQUEUE *)

   IF Q.FRONT = NIL THEN
      WRITELN( '<empty>' )
   ELSE
      BEGIN (* printing *)
      P := Q.FRONT;
      REPEAT
         WRITELN( P@.INFO );
         P := P@.NEXT
      UNTIL P = NIL
      END (* printing *)

   END; (* PRINTQUEUE *)
```

†*Ex. 3: A queue can be implemented without an extra pointer to the front: have a pointer to the rear node, and make that node always point to the front one instead of always being NIL. This implementation is a circular queue. Draw a diagram of an example circular queue, show the declarations, and write appropriate ENQUEUE and REMOVE procedures.*

LINKED LISTS IN GENERAL

The greatest difference between a stack and a queue is the way in which each is used; they *look* much the same. They are in fact special instances of linear linked lists. These lists are linear, homogeneous data structures, as are one-dimensional arrays (called simply "lists") and files. Linked lists are stored in memory as arrays are, but are sequential and unbounded like files.

Because arrays, files, and linear linked lists are so similar in structure, it is not surprising that many of the same operations can be performed on linked lists as on arrays and files.

Assume that we have declared a general linear linked list type:

```
TYPE
    LINKLIST =
        RECORD
        HEAD : NODEPTR;      (* Pointer to first node in list. *)
        SENTINEL : NODEPTR   (* Pointer to a sentinel node. *)
        END;

VAR
    LIST : LINKLIST;
    ITEM : INFOTYPE;
```

As before, LIST can contain zero or more nodes of information. This time, however, there is a significant difference:

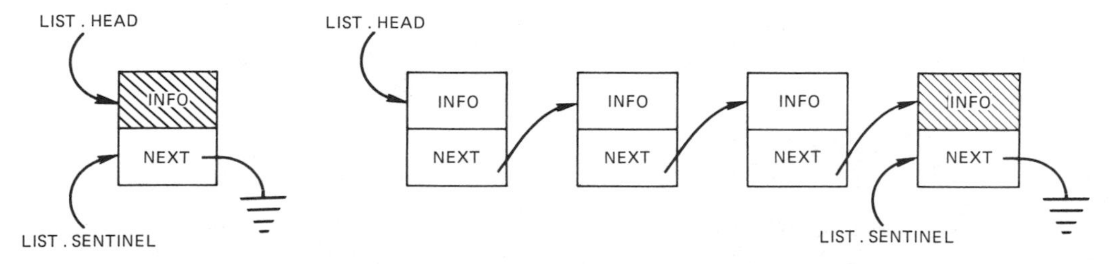

Here LIST.HEAD points to the first node in the list, while LIST.SENTINEL points to an ever-present sentinel node whose INFO field contains a "dummy" (unused) value. An "empty" list contains only the sentinel. As you saw in the discussions about standard input, character processing, and arrays, a sentinel at the end of a list can make algorithms easier.

The procedure to create an empty list must set up the sentinel node:

```
PROCEDURE  LISTINIT( VAR LIST : LINKLIST );
   (* Initialize the given LIST to "empty"--that is,         *)
   (* containing only a sentinel node.                       *)

   BEGIN (* LISTINIT *)

   WITH LIST DO
      BEGIN
      NEW( SENTINEL );
      HEAD := SENTINEL;
      SENTINEL@.INFO := ...; (* A dummy value.               *)
      SENTINEL@.NEXT := NIL
      END
   END; (* LISTINIT *)
```

The list is empty if the sentinel is the only node:

```
FUNCTION  LISTEMPTY( LIST : LINKLIST ) : BOOLEAN;
   (*Return TRUE if the given LIST is empty.                 *)

   BEGIN (* LISTEMPTY *)

   LISTEMPTY := ( LIST.HEAD = LIST.SENTINEL )

   END; (* LISTEMPTY *)
```

Input and output are straightforward. Assume that we wish to store values in the same order in which they occur in the input stream:

```
PROCEDURE  GETLIST( VAR LIST : LINKLIST );
   (* Read and store input values into the given list.      *)

   VAR
      PREV : NODEPTR;    (* Points to previous node.         *)
      CURR : NODEPTR;    (* Points to current node.          *)

   BEGIN (* GETLIST *)
   LISTINIT( LIST );

   (* Get and save the input values until EOF:               *)
   WHILE NOT EOF(INPUT) DO
      BEGIN (* each value *)
      NEW( CURR );
      READLN( CURR@.INFO );
      CURR@.NEXT := LIST.SENTINEL;
      IF LIST.HEAD = LIST.SENTINEL THEN
         LIST.HEAD := CURR
      ELSE
         PREV@.NEXT := CURR;
      PREV := CURR
      END (* each value *)
   END; (* GETLIST *)
```

Output is simpler:

```
PROCEDURE  PRINTLIST( LIST : LINKLIST );
   (* Print the contents, if any, of the given list.      *)

   VAR
      P : NODEPTR;         (* For stepping through the list. *)

   BEGIN (* PRINTLIST *)

   WITH LIST DO
      IF HEAD = SENTINEL THEN
         WRITELN( '<empty>' )
      ELSE
         BEGIN (* printing *)
         P := HEAD;
         REPEAT
            WRITELN( P@.INFO );
            P := P@.NEXT
         UNTIL P = SENTINEL
         END (*printing *)

   END; (* PRINTLIST *)
```

Ex. 4: Write a function to sum the contents of a linked list, assuming that INFOTYPE is INTEGER.

†*Ex. 5: Redefine the LINKLIST type without a sentinel node but with a LAST pointer to the last node of data. Rewrite LISTINIT, LISTEMPTY, and PRINTLIST for this implementation.*

The values stored in the list are typically (but not necessarily) ordered. Assuming that they are in ascending order from HEAD to SENTINEL, the list can be searched to see where the value of ITEM ought to be found. This function returns a pointer to that location:

```
FUNCTION  SEARCH( LIST: LINKLIST; ITEM: INFOTYPE ): NODEPTR;
   (* Return the position in LIST where the value of ITEM *)
   (* is or should be.  If the value would be beyond the  *)
   (* end of the list, return the sentinel pointer.       *)

   VAR
      P : NODEPTR;         (* For stepping through the list. *)

   BEGIN (* SEARCH *)

   P := LIST.HEAD;
   WHILE (P <> LIST.SENTINEL) AND (P@.INFO < ITEM) DO
      P := P@.NEXT;
   SEARCH := P

   END; (* SEARCH *)
```

Ex. 6: Rewrite SEARCH using ITEM.INFO as the sentinel node's value, thus simplifying the loop.

If we have found where a given value ought to be in LIST, it can be put there this way:

```
PROCEDURE  INSERT( ITEM : INFOTYPE; VAR LIST : LINKLIST;
                                    VAR POS : NODEPTR );
   (* Insert a node containing the value of ITEM into      *)
   (* LIST just before the node at position POS.           *)
   VAR
      NEWNODE : NODEPTR; (* Pointer to the new node.       *)

   BEGIN (* INSERT *)

   NEW( NEWNODE );

   (* Put the current values at POS into the new node:     *)
   NEWNODE@ := POS@;

   (* Link the node at POS to be prior to this one:        *)
   POS@.NEXT := NEWNODE;

   (* Put the new value into the node at POS, which is     *)
   (* now the node prior to the old information:           *)
   POS@.INFO := ITEM;

   (* If POS was the sentinel node then reset the latter: *)
   IF POS = LIST.SENTINEL THEN
      LIST.SENTINEL := NEWNODE;

   (* In any case, POS must point to its original data:    *)
   POS := NEWNODE

   END; (* INSERT *)
```

Since it is impossible to insert a new node prior to a given node if only the position of that given node is known, the new node was inserted *after* the given one, then the contents of the two nodes were swapped so that the information is now in the correct order.

Ex. 7: Rewrite GETLIST in terms of INSERT.

†*Ex. 8: Write an algorithm, in terms of these procedures, to build an ordered list from unordered input.*

†*Ex. 9: Write an algorithm to order an unordered list. (There are several ways to do this, including the equivalents of bubble, selection, and insertion sorts.)*

Ex. 10: Write a procedure called REINIT that disposes of any nodes currently allocated to a given list, then initializes the list to empty. Why is this procedure not interchangeable with LISTINIT?

Assume that we have searched for and found a given value and now wish to remove it from the list. This can be done using an algorithm that swaps node contents again:

```
PROCEDURE  DELETE( VAR POS : NODEPTR; VAR LIST : LINKLIST );
   (* Remove the node at position POS from the given LIST.*)
   (* Upon return, the actual parameter for POS has the   *)
   (* value NIL.                                           *)

   VAR
      VICTIM : NODEPTR; (* Pointer to the node to delete. *)

   BEGIN (* DELETE *)

   VICTIM := POS@.NEXT;
   POS@ := VICTIM@;
   IF VICTIM = LIST.SENTINEL THEN
      LIST.SENTINEL := POS;
   POS := NIL;
   DISPOSE( VICTIM )
   END; (* DELETE *)
```

The last of the basic algorithms for general linked lists is one to compare two lists for equality:

```
FUNCTION  EQUAL( L1 : LINKLIST; L2 : LINKLIST ) : BOOLEAN;
   (* Return TRUE if the two given lists have equal        *)
   (* values in their corresponding nodes.                 *)

   VAR
      P : NODEPTR;         (* For stepping through list L1.  *)
      Q : NODEPTR;         (* For stepping through list L2.  *)
      MATCH : BOOLEAN;     (* TRUE while the two lists match.*)

   BEGIN (* EQUAL *)

   (* Step through the two lists in parallel, comparing   *)
   (* values in the nodes until a mismatch or the end of  *)
   (* a list is found.                                     *)
   P := L1.HEAD;
   Q := L2.HEAD;
   MATCH := TRUE;
   WHILE (P <> L1.SENTINEL) AND (Q <> L2.SENTINEL)
         AND MATCH DO
      IF P@.INFO = Q@.INFO THEN
         BEGIN (* step *)
         P := P@.NEXT;
         Q := Q@.NEXT
         END (* step *)
      ELSE
         MATCH := FALSE;

   (* To be equal, the lists must also be the same length:*)
   EQUAL :=MATCH AND (P = L1.SENTINEL) AND (Q = L2.SENTINEL)
   END; (* EQUAL *)
```

†*Ex. 11: Rewrite SEARCH, INSERT, and DELETE using a definition of LINKLIST without a sentinel node but with a LAST pointer to the last node containing information (as in an earlier exercise).*

These basic algorithms can now be used to implement some more complex ones. For example, to catenate the contents of two lists, producing a third:

```
PROCEDURE  CATLIST( L1,L2 : LINKLIST; VAR L3 : LINKLIST );
   (* Copy the contents of L1 followed by L2 into L3.     *)

   VAR
      P : NODEPTR;        (* For stepping through L1 and L2.*)

   BEGIN (* CATLIST *)

   (* Assume that list L3 has never been initialized:     *)
   LISTINIT( L3 );

   (* Copy L1 to L3 by inserting the value of each node   *)
   (* of L1 into L3 just prior to L3's sentinel:          *)
   P := L1.HEAD;
   WHILE P <> L1.SENTINEL DO
      INSERT( P@.INFO, L3, L3.SENTINEL );

   (* Copy L2's values to L3 the same way:                *)
   P := L2.HEAD;
   WHILE P <> L2.SENTINEL DO
      INSERT( P@.INFO, L3, L3.SENTINEL )

   END; (* CATLIST *)
```

Notice that in linked-list algorithms there is an important distinction between assigning the *actual nodes* from one list to another and assigning *copies of node values* to new nodes created to hold those values.

There is another important and unusual property of pointers. When the value of a value parameter is changed inside a procedure, the actual parameter with which the procedure was called is not affected. This is also true of pointers passed as value parameters. However, *the objects pointed to can be changed* from inside the procedure!

Ex. 12: Write a catenation procedure that appends the actual nodes of L2 to L1 and then initializes L2 to empty.

As was done with arrays, strings, and files, an ordered linked list can be compressed by deleting duplicate entries:

```
PROCEDURE  COMPRESS( VAR LIST : LINKLIST );
   (* Compress LIST by removing duplicate entries.  LIST  *)
   (* is ordered.                                         *)

   VAR
      FORE : NODEPTR;      (* For stepping through the list. *)
      AFT : NODEPTR;       (* Trails one node behind FORE.   *)

   BEGIN (* COMPRESS *)

   (* Point AFT and FORE to the first and second nodes:   *)
   AFT := LIST.HEAD;
   FORE := LIST.HEAD;
   IF FORE <> LIST.SENTINEL THEN
      FORE := FORE@.NEXT;

   (* Step through the list, comparing the values in      *)
   (* consecutive nodes, removing the first of any two    *)
   (* matching nodes:                                     *)
   WHILE FORE <> LIST.SENTINEL DO
      BEGIN (* each pair *)
      IF AFT@.INFO = FORE@.INFO THEN
         BEGIN (* remove duplicate *)
         DELETE( FORE, LIST );
         FORE := AFT@.NEXT
         END (* remove duplicate *)
      ELSE
         BEGIN (* step along *)
         AFT := FORE;
         FORE := FORE@.NEXT
         END (* step along *)
      END (* each pair *)

   END; (* COMPRESS *)
```

†*Ex. 13: Write a procedure that takes one list and returns a new list that contains the values of the first without duplicates. (The given list is ordered.)*

We conclude this section with an algorithm that merges two ordered lists to produce a new ordered list:

```
PROCEDURE  MERGE( L1,L2 : LINKLIST; VAR L3 : LINKLIST );
   (* Merge the contents of the ordered lists L1 and L2    *)
   (* to form a new ordered list L3.                        *)

   VAR
      P : NODEPTR;         (* For stepping through L1.      *)
      Q : NODEPTR;         (* For stepping through L2.      *)

   PROCEDURE SAVENSTEP( VAR P : NODEPTR; VAR L : LINKLIST );
      (* Save the INFO at P into L before its sentinel,     *)
      (* then step P to the next node in its list.          *)

      BEGIN (* SAVENSTEP *)
      INSERT( P@.INFO, L, L.SENTINEL );
      P := P@.NEXT
      END; (* SAVENSTEP *)

   BEGIN (* MERGE *)

   (* Assume a completely new list L3:                      *)
   LISTINIT( L3 );

   (* Step through the two lists at once, always putting   *)
   (* the smaller of the two INFO values into L3:          *)
   P := L1.HEAD;
   Q := L2.HEAD;
   WHILE (P <> L1.SENTINEL) AND (Q <> L2.SENTINEL) DO
      IF P@.INFO < Q@.INFO THEN
         SAVENSTEP( P, L3 )
      ELSE
         SAVENSTEP( Q, L3 );

   (* One of the two lists may not be finished yet:        *)
   WHILE P <> L1.SENTINEL DO
      SAVENSTEP( P, L3 );
   WHILE Q <> L2.SENTINEL DO
      SAVENSTEP( Q, L3 )

   END; (* MERGE *)
```

Ex. 14: There are two other ways to do a merge:
1. Incorporate the actual nodes of one list into the other.
2. Incorporate copies of values from one list as new nodes in the other.
Write and compare these algorithms.

Ex. 15: Rewrite the original merge algorithm, assuming that linked lists do not have sentinel nodes.

BINARY TREES

A binary tree is a nonlinear linked structure wherein each node can link to as many as two other nodes. The binary *search* trees discussed in this section can hold any quantity of information and can be accessed quickly in *sorted order.* These trees are therefore frequently used instead of arrays, files, or linear linked lists to hold ordered sets of information.

A tree can be defined with these declarations:

```
TYPE
   INFOTYPE = ...;           (* Assume a simple type for now.   *)
   TREEPTR = @TREENODE;
   TREENODE =                (* A node with data and two links.*)
      RECORD
      INFO : INFOTYPE;
      LEFT : TREEPTR;        (* To nodes with INFO < this one. *)
      RIGHT : TREEPTR        (* To nodes with INFO > this one. *)
      END;
   TREE =                    (* All variables of type TREE     *)
      RECORD                 (* will have a ROOT:              *)
      ROOT : TREEPTR         (* Points to first node in tree.  *)
      END;

VAR
   BINTREE : TREE;
```

A binary tree may include zero or more nodes. Examples are:

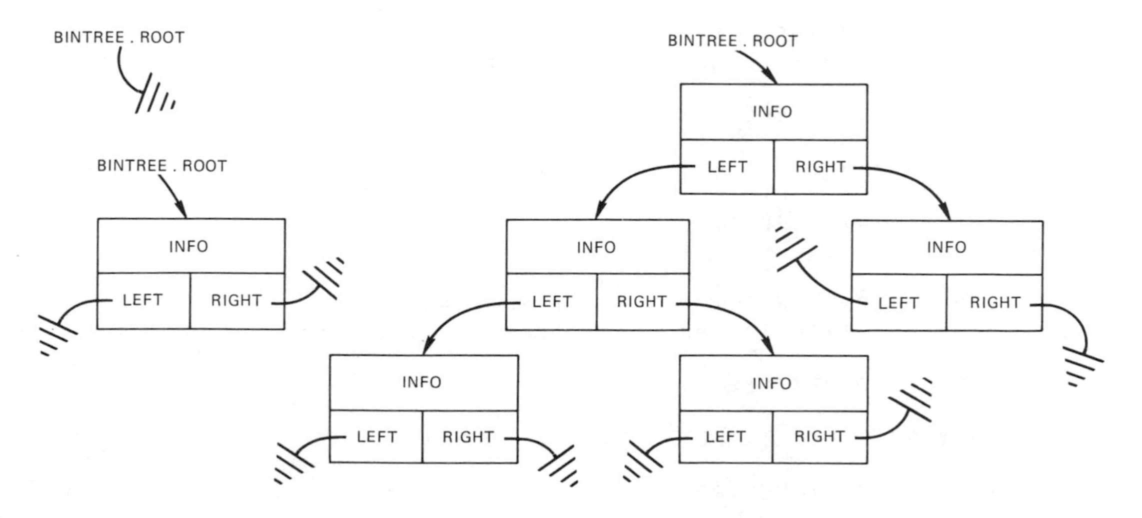

The information in a binary search tree is stored such that all values smaller than the one in a given node are in that node's left subtree; all larger values are in its right subtree. (The subtrees of a node are just the smaller tree structures to which the node points.) For example, assuming that the INFO fields are integers:

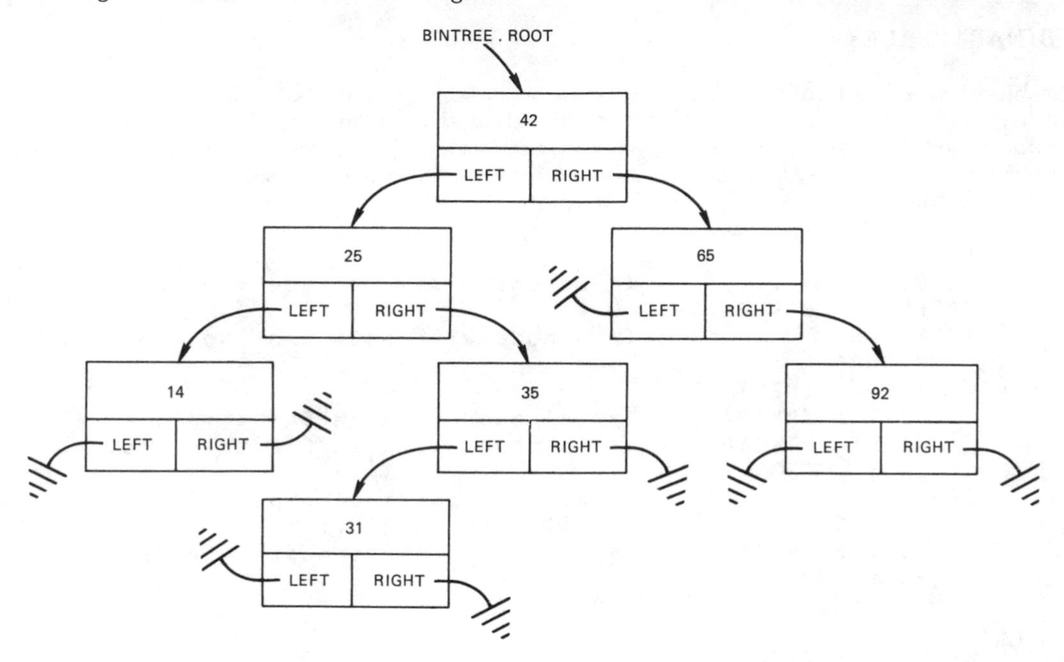

The simple procedures, those to initialize a tree to empty and to see if one has become empty, are:

```
PROCEDURE  INITTREE( VAR T : TREE );
   (* Initialize a binary tree to empty.                    *)

   BEGIN (* INITTREE *)

   T.ROOT := NIL

   END; (* INITTREE *)


FUNCTION  EMPTYTREE( T : TREE ): BOOLEAN;
   (* Return TRUE if the given tree is empty.               *)

   BEGIN (* EMPTYTREE *)

   EMPTYTREE := ( T.ROOT = NIL )

   END; (* EMPTYTREE *)
```

It is also convenient to have a procedure that creates a new node to contain some given information:

```
FUNCTION  NEWNODE( ITEM : INFOTYPE ) : TREEPTR;
   (* Return a pointer to a node containing the value of  *)
   (* the given item and initialized pointers.            *)

   VAR
      P : TREEPTR;        (* A temporary pointer.          *)

   BEGIN (* NEWNODE *)

   NEW( P );
   WITH P@ DO
      BEGIN
      INFO := ITEM;
      LEFT := NIL;
      RIGHT := NIL
      END;
   NEWNODE := P

   END; (* NEWNODE *)
```

Given a tree, a search procedure can be used to see if it already contains a given value:

```
FUNCTION  SEARCH( T : TREE; ITEM : INFOTYPE ): TREEPTR;
   (* Return a pointer to the node containing the value   *)
   (* of ITEM, if any, or a NIL pointer otherwise.         *)

   VAR
      P : TREEPTR;        (* For stepping through the tree. *)
      FOUND : BOOLEAN;    (* Becomes TRUE when value found. *)

   BEGIN (* SEARCH *)

   (* Starting at the ROOT node, follow the links in the  *)
   (* direction of the given value:                        *)
   P := T.ROOT;
   FOUND := FALSE;
   WHILE (P <> NIL) AND NOT FOUND DO
      IF ITEM = P@.INFO THEN
         FOUND := TRUE
      ELSE IF ITEM < P@.INFO THEN
         P := P@.LEFT
      ELSE
         P := P@.RIGHT;

   (* Now P either points to the value, or P has "stepped *)
   (* off" a node and is now NIL:                          *)
   SEARCH := P

   END; (* SEARCH *)
```

To insert a new value into a tree, SEARCH can not be called to find out where to put it, since SEARCH may return a nil value. We must use a procedure that contains its own peculiar search:

```
PROCEDURE  INSERT( ITEM : INFOTYPE; VAR T : TREE );
   (* Insert the value of ITEM into the given tree T, if  *)
   (* it is not already there: (do nothing otherwise)     *)

   VAR
      P : TREEPTR;          (* For stepping through the tree. *)
      Q : TREEPTR;          (* Trails one node behind P.      *)
      FOUND : BOOLEAN;      (* Becomes TRUE if value is found.*)

   BEGIN (* INSERT *)

   (* Search T for the value of ITEM, keeping Q one node   *)
   (* behind P so the last node inspected is always known:*)
   P := T.ROOT;
   Q := NIL;
   FOUND := FALSE;
   WHILE (P <> NIL) AND NOT FOUND DO
      IF ITEM = P@.INFO THEN
         FOUND := TRUE
      ELSE
         BEGIN (* keep looking *)
         Q := P;
         IF ITEM < P@.INFO THEN
            P := P@.LEFT
         ELSE
            P := P@.RIGHT
         END; (* keep looking *)

   (* Now insert a new node for the value, if necessary:  *)
   IF NOT FOUND THEN
      BEGIN (* create one *)
      P := NEWNODE( ITEM );
      IF Q = NIL THEN
         (* The tree was empty:                            *)
         T.ROOT := P
      ELSE
         (* The node at Q must be linked to the new node: *)
         IF ITEM < Q@.INFO THEN
            Q@.LEFT := P
         ELSE
            Q@.RIGHT := P
      END (* create one *)

   END; (* INSERT *)
```

INSERT is often written as a procedure that returns a pointer to the node containing the given information. It thus acts like a search that is always successful, inserting when necessary.

To print the contents of a binary search tree in sorted order, we need a simple recursive procedure:

```
PROCEDURE  PRINTTREE( T : TREE );
   (* Calls PRINTNODES to print the contents, if any, of  *)
   (* the given tree.                                      *)

   PROCEDURE  PRINTNODES( P : TREEPTR );
      (* Recursively traverses the nodes starting at P,   *)
      (* printing them in ascending order.                *)

      BEGIN (* PRINTNODES *)

      (* Print all values less than this one, if any:     *)
      IF P@.LEFT <> NIL THEN
         PRINTNODES( P@.LEFT );

      (* Then print this one:                             *)
      WRITELN( P@.INFO );

      (* Then print all values greater than this, if any: *)
      IF P@.RIGHT <> NIL THEN
         PRINTNODES( P@.RIGHT )

      END; (* PRINTNODES *)

   BEGIN (* PRINTTREE *)

   IF T.ROOT = NIL THEN
      WRITELN( '<empty>' )
   ELSE
      PRINTNODES( T.ROOT )

   END; (* PRINTTREE *)
```

†*Ex. 16: Rethink, then rewrite, PRINTNODES so that it needs to make only one test.*

PROGRAM FOR READING

```
 1  PROGRAM  CROSSREF( INPUT, OUTPUT );
 2
 3  (* PURPOSE:                                                    *)
 4  (*    Print a given text and a cross-reference, by line        *)
 5  (*    numbers, of all the words in it.                         *)
 6  (*                                                             *)
 7  (* PROGRAMMER:  David V. Moffat                                *)
 8  (*                                                             *)
 9  (* INPUT:                                                      *)
10  (*    Any textual material, from the file INPUT.               *)
11  (*                                                             *)
12  (* OUTPUT:                                                     *)
13  (*    An echo of the input text, with its lines numbered,      *)
14  (*    and a listing of each word along with the lines in       *)
15  (*    the text on which it occurs.                             *)
16  (*                                                             *)
17  (* ASSUMPTIONS & LIMITATIONS:                                  *)
18  (*    All punctuation will be removed.  Hyphenated words       *)
19  (*       and contractions are not handled properly.            *)
20  (*    Capitalized words are uncapitalized.                     *)
21  (*    Words may contain up to 20 characters.                   *)
22  (*                                                             *)
23  (* ALGORITHM:                                                  *)
24  (*    Initialize the storage structure and print titles.       *)
25  (*    Process the text until EOF:                              *)
26  (*       Skip the characters between words.                    *)
27  (*       Scan for and accumulate the next word.                *)
28  (*       Store it with its reference.                          *)
29  (*    Print the cross-reference table.                         *)
30
```

```
31  CONST
32     MAXSTRING = 20;          (* Max. string (word) length.   *)
33
34  TYPE
35     FIXEDSTRING = PACKED ARRAY[1..MAXSTRING] OF CHAR;
36     STRING =                 (* "Varying-length" strings:    *)
37        RECORD
38        STR : FIXEDSTRING;
39        LEN : 0..MAXSTRING
40        END;
41
42     NODEPTR = @LISTNODE;     (* Linked list of line numbers:*)
43     LISTNODE =
44        RECORD
45        REF : INTEGER;        (* A line (reference) number.   *)
46        NEXT : NODEPTR
47        END;
48     LINKLIST =
49        RECORD
50        HEAD : NODEPTR;
51        LAST : NODEPTR
52        END;
53
54     TREEPTR = @TREENODE;     (* Binary search tree of words:*)
55     TREENODE =
56        RECORD
57        WORD : STRING;        (* One word from the text.      *)
58        REFS : LINKLIST;      (* Lines upon which it's found.*)
59        LEFT : TREEPTR;
60        RIGHT : TREEPTR
61        END;
62     TREE =
63        RECORD
64        ROOT : TREEPTR
65        END;
66
67     CLASSES =                (* Character classes of text:   *)
68        (BLANK,               (* The blank.                   *)
69         LETTER,              (* All alphabetics.             *)
70         OTHER,               (* All other characters.        *)
71         ENDFILE);            (* End of file flag.            *)
72
73  VAR
74     LINE : INTEGER;          (* Current line number.         *)
75     REFLINE : INTEGER;       (* Line number THISWORD is on.  *)
76     THISWORD : STRING;       (* Next word from the text.     *)
77     STORE : TREE;            (* Storage area to hold words.  *)
78     CH : CHAR;               (* One character of input text.*)
79     CLASS : CLASSES;         (* Class of the input char.     *)
80     GOTAWORD : BOOLEAN;      (* TRUE when word accumulated.  *)
81
```

```
 82  PROCEDURE  GETNEXT( VAR CH : CHAR; VAR CLASS : CLASSES;
 83                      VAR LINE : INTEGER );
 84     (* Get, echo, and return the next input char, CH, its  *)
 85     (* character CLASS, and the LINE number on which it is *)
 86     (* found.  All alphabetics are returned in lowercase.  *)
 87     (* CH is set to BLANK upon EOLN.                        *)
 88
 89     BEGIN (* GETNEXT *)
 90
 91     (* Check the special cases first, then try to get CH:  *)
 92     IF EOF(INPUT) THEN
 93        CLASS := ENDFILE
 94     ELSE IF EOLN(INPUT) THEN
 95        BEGIN (* line end *)
 96        WRITELN;
 97        LINE := LINE + 1;
 98        WRITE( LINE:3, '> ' );
 99        READ( CH );
100        CLASS := BLANK
101        END (* line end *)
102     ELSE
103        BEGIN (* process next char *)
104        READ( CH );
105        WRITE( CH );
106        (* Translate uppercase into lowercase:              *)
107        IF CH IN ['A'..'Z'] THEN
108           CH := CHR( ORD(CH)-ORD('A')+ORD('a') );
109        (* Determine the character's class:                 *)
110        IF CH IN ['a'..'z'] THEN
111           CLASS := LETTER
112        ELSE IF CH = ' ' THEN
113           CLASS := BLANK
114        ELSE
115           CLASS := OTHER
116        END (* process next char *)
117     END; (* GETNEXT *)
118
119
120  PROCEDURE  CLEAR( VAR WORD : STRING );
121     (* Set the given WORD to the "null" string.             *)
122
123     BEGIN (* CLEAR *)
124     WORD.STR := '                    ';
125     WORD.LEN := 0
126     END; (* CLEAR *)
127
128
129  PROCEDURE  INITLIST( VAR L : LINKLIST );
130     (* Initialize the given linked list, L, to empty.      *)
131
132     BEGIN (* INITLIST *)
133     L.HEAD := NIL;
134     L.LAST := NIL
135     END; (* INITLIST *)
136
```

```
137   PROCEDURE  CATCHAR( VAR WORD : STRING; CH : CHAR );
138      (* Catenate the given char, CH, to the right end of    *)
139      (* the given WORD, if possible.                         *)
140
141      BEGIN (* CATCHAR *)
142      WITH WORD DO
143         IF LEN < MAXSTRING THEN
144            BEGIN (* save char *)
145            LEN := LEN + 1;
146            STR[LEN] := CH
147            END (* save char *)
148      END; (* CATCHAR *)
149
150
151   FUNCTION  MAKENODE( WORD : STRING ) : TREEPTR;
152      (* Create, initialize, and return a TREEPTR to a node  *)
153      (* containing the given WORD and its reference list.   *)
154
155      VAR
156         P : TREEPTR;            (* Temporary pointer.          *)
157
158      BEGIN (* MAKENODE *)
159      NEW( P );
160      CLEAR( P@.WORD );
161      P@.WORD := WORD;
162      INITLIST( P@.REFS );
163      P@.LEFT := NIL;
164      P@.RIGHT := NIL;
165      MAKENODE := P
166      END; (* MAKENODE *)
167
168
169   PROCEDURE  INSERTLIST( VAR L : LINKLIST; LINE : INTEGER );
170      (* Insert into L a new node containing the LINE number.*)
171
172      VAR
173         P : NODEPTR;            (* Temporary pointer.          *)
174
175      BEGIN (* INSERTLIST *)
176      NEW( P );
177      P@.REF := LINE;
178      P@.NEXT := NIL;
179      WITH L DO
180         IF HEAD = NIL THEN
181            BEGIN
182            HEAD := P;
183            LAST := P
184            END
185         ELSE
186            BEGIN
187            LAST@.NEXT := P;
188            LAST := P
189            END
190      END; (* INSERTLIST *)
191
```

```
192  PROCEDURE  INSERT( ITEM : STRING; LINE : INTEGER;
193                     VAR T : TREE );
194     (* Insert the value of ITEM into the given tree T, if  *)
195     (* it is not already there.  In any case, include the  *)
196     (* given LINE number in its list of references.        *)
197
198     VAR
199        P : TREEPTR;        (* For stepping through the tree. *)
200        Q : TREEPTR;        (* Trails one node behind P.      *)
201        FOUND : BOOLEAN;    (* Becomes TRUE if value is found.*)
202
203     BEGIN (* INSERT *)
204
205     (* Search T for the value of ITEM, keeping Q one node  *)
206     (* behind P so the last node inspected is always known:*)
207     P := T.ROOT;
208     Q := NIL;
209     FOUND := FALSE;
210     WHILE (P <> NIL) AND NOT FOUND DO
211        IF ITEM.STR = P@.WORD.STR THEN
212           FOUND := TRUE
213        ELSE
214           BEGIN (* keep looking *)
215           Q := P;
216           IF ITEM.STR < P@.WORD.STR THEN
217              P := P@.LEFT
218           ELSE
219              P := P@.RIGHT
220           END; (* keep looking *)
221
222     (* Now insert a new node for the value, if necessary:  *)
223     IF NOT FOUND THEN
224        BEGIN (* create one *)
225        P := MAKENODE( ITEM );
226        IF Q = NIL THEN
227           (* The tree was empty:                            *)
228           T.ROOT := P
229        ELSE
230           (* The node at Q must link to the new node:       *)
231           IF ITEM.STR < Q@.WORD.STR THEN
232              Q@.LEFT := P
233           ELSE
234              Q@.RIGHT := P
235        END; (* create one *)
236
237     (* Include the given line number in the references:    *)
238     INSERTLIST( P@.REFS, LINE )
239
240     END; (* INSERT *)
241
```

```
242  PROCEDURE  PRINT( T : TREE );
243     (* Prints the contents of T in ascending order by word.*)
244
245     PROCEDURE  PRINTLIST( L : LINKLIST );
246        (* Prints the reference numbers in L, 10 per line.  *)
247
248        VAR
249           P : NODEPTR;    (* For stepping through the list. *)
250           N : INTEGER;    (* Number of references on a line.*)
251
252        BEGIN (* PRINTLIST *)
253        P := L.HEAD;
254        N := 0;
255        REPEAT
256           N := N + 1;
257           WRITE( P@.REF:4 );
258           (* Need a new indented line if this one is full: *)
259           IF (N = 10) AND (P <> L.LAST) THEN
260              BEGIN (* new line *)
261              WRITELN;
262              N := 0;
263              WRITE( ' ':25 )
264              END; (* new line *)
265           P := P@.NEXT
266        UNTIL P = NIL;
267        WRITELN
268        END; (* PRINTLIST *)
269
270     PROCEDURE  PRINTWORDS( P : TREEPTR );
271        (* Recursive inorder traversal to print the tree    *)
272        (* pointed to by P.                                 *)
273
274        BEGIN (* PRINTWORDS *)
275        IF P <> NIL THEN
276           WITH P@ DO
277              BEGIN (* print all *)
278              PRINTWORDS( LEFT );
279              WRITE( WORD.STR, ' ':5 );
280              PRINTLIST( REFS );
281              PRINTWORDS( RIGHT )
282              END (* print all *)
283        END; (* PRINTWORDS *)
284
285     BEGIN (* PRINT *)
286
287     (* Print titles for the table of references:           *)
288     WRITELN( 'Word Cross-reference:':30 );
289     WRITELN;
290     WRITELN( 'Word:', ' ':20, 'References:' );
291
292     (* Print the table:                                    *)
293     PRINTWORDS( T.ROOT )
294
295     END; (* PRINT *)
296
```

```
297  BEGIN (* CROSSREF *)
298
299  (* Print titles for the output:                                  *)
300  PAGE( OUTPUT );
301  WRITELN( 'Text to be cross-referenced:' );
302  WRITELN;
303  LINE := 1;
304  WRITE( LINE:3, '> ' );
305
306  (* Initialize the structure for storing the words:               *)
307  STORE.ROOT := NIL;
308
309  (* Process the text a character at a time until EOF,             *)
310  (* accumulating words and saving them in STORE along with *)
311  (* their line numbers:                                            *)
312  CLEAR( THISWORD );
313  GETNEXT( CH, CLASS, LINE );
314  REPEAT
315
316     (* Skip the characters between words:                         *)
317     WHILE CLASS IN [BLANK,OTHER] DO
318        GETNEXT( CH, CLASS, LINE );
319
320     (* Scan the word, if any, accumulating its characters: *)
321     GOTAWORD := FALSE;
322     REFLINE := LINE;
323     WHILE CLASS IN [LETTER] DO
324        BEGIN (* each char *)
325        GOTAWORD := TRUE;
326        CATCHAR( THISWORD, CH );
327        GETNEXT( CH, CLASS, LINE )
328        END;
329
330     (* If a word was encountered, save it and its line #:  *)
331     IF GOTAWORD THEN
332        BEGIN
333        INSERT( THISWORD, REFLINE, STORE );
334        CLEAR( THISWORD )
335        END
336
337  UNTIL CLASS = ENDFILE;
338
339  WRITELN( '***End of text.' );
340
341  (* Print the words with their references in alphabetical  *)
342  (* order:                                                         *)
343  PAGE( OUTPUT );
344  PRINT( STORE );
345  WRITELN;
346  WRITELN( 'End of program.' );
347  PAGE( OUTPUT )
348
349  END. (* CROSSREF *)
```

```
Text to be cross-referenced:

   1> PROCEDURE  PRINT( T : TREE );
   2>
   3>    PROCEDURE  PRINTLIST( L : LINKLIST );
   4>       VAR
   5>          P : NODEPTR;
   6>          N : INTEGER;
   7>
   8>       BEGIN
   9>       P := L.HEAD;
  10>       N := 0;
  11>       REPEAT
  12>          N := N + 1;
  13>          WRITE( P@.REF:4 );
  14>          IF (N = 10) AND (P <> L.LAST) THEN
  15>             BEGIN
  16>             WRITELN;
  17>             N := 0;
  18>             WRITE( ' ':25 )
  19>             END;
  20>          P := P@.NEXT
  21>       UNTIL P = NIL;
  22>       WRITELN
  23>       END;
  24>
  25>    PROCEDURE  PRINTWORDS( P : TREEPTR );
  26>       BEGIN
  27>       IF P <> NIL THEN
  28>          BEGIN
  29>          PRINTWORDS( P@.LEFT );
  30>          WRITE( P@.WORD.STR, ' ':5 );
  31>          PRINTLIST( P@.REFS );
  32>          PRINTWORDS( P@.RIGHT )
  33>          END
  34>       END;
  35>
  36>
  37>    BEGIN
  38>    WRITELN;
  39>    PRINTWORDS( T.ROOT )
  40>    END;
  41> ***End of text.
```

```
         Word Cross-reference:

Word:                      References:
and                          14
begin                         8   15   26   28   37
end                          19   23   33   34   40
head                          9
if                           14   27
integer                       6
l                             3    9   14
last                         14
left                         29
linklist                      3
n                             6   10   12   12   14   17
next                         20
nil                          21   27
nodeptr                       5
p                             5    9   13   14   20   20   21   25   27   29
                             30   31   32
print                         1
printlist                     3   31
printwords                   25   29   32   39
procedure                     1    3   25
ref                          13
refs                         31
repeat                       11
right                        32
root                         39
str                          30
t                             1   39
then                         14   27
tree                          1
treeptr                      25
until                        21
var                           4
word                         30
write                        13   18   30
writeln                      16   22   38

End of program.
```

PROGRAM EXERCISES

1. (a) How do the definitions of the LINKLIST type and the INITLIST, INSERTLIST, and PRINTLIST procedures differ from those discussed earlier?
   (b) How would such a linked list be drawn?
   (c) Would the use of the previously discussed linked-list implementation change the organization of this program?

2. Note that, even though CROSSREF has a GETNEXT procedure, as described in Part II, the main loop of the program uses a sentinel strategy rather than a transition strategy. Which sections of the program would have to be changed if a character transition strategy were to be used instead?

3. The body of the program contains one statement that depends upon the implementation details of the major data structure. Find it, then describe how that dependence can be avoided.

4. Find everything that would have to be changed if STORE were to be changed to a linear linked list. Describe these changes. In particular, are more, fewer, or the same number of procedures needed?

Part VIII: Special Topics

INTRODUCTION

This part presents several algorithms and strategies that are sometimes thought to be "missing" from Pascal by persons who already program in another language.

The first topic, information hiding, is of importance to all programmers because it offers strategies and a discipline that contribute to the quality of any program.

The need for some kind of formatted input is especially urgent to programmers who are accustomed to using it in other languages. Pascal programmers will sometimes find it a convenient way to simplify input procedures.

Exact-precision fractions can be important in applications that deal with monetary values, particularly when using versions of Pascal for microcomputers whose real types have few significant digits.

Procedures for generating pseudorandom numeric sequences are provided here because they are difficult to find elsewhere.

INFORMATION HIDING

Information hiding refers to methods by which the details of the implementation of program operations and data types are "hidden" from the rest of the program. This is done to protect implementation details from accidental changes, to free the rest of the program from dependence upon specific details of the implementation (thus making the program easier to modify), and to promote clear program structure. It also goes hand in hand with top-down program development.

Information hiding is promoted by specific language features and by programming methods and style. The single most important language feature in support of information hiding is the UNIT feature of UCSD Pascal. The next best mechanism is the "separate compilation" of procedures and functions. Since neither of these language features is a part of standard Pascal (although they are *usually* in extended Pascals), we will not discuss them here. (See the Bibliography for references to UCSD Pascal.) Instead, we will present five programming strategies that contribute to information hiding.

The first strategy is simply to avoid using nested procedures, because they inherit variables from surrounding scopes, and so can inadvertently use or alter those variables. If we want to take this idea to its limit, we can organize a program this way:

```
PROGRAM  DISINHERIT( INPUT, OUTPUT );
Declarations of global constants and types only.

PROCEDURE  A( ... );
   Declarations of local variables.
   BEGIN (* A *)
   ...
   END; (* A *)

PROCEDURE  B( ... );
   Local variables.
   BEGIN (* B *)
   ...
   END; (* B *)

PROCEDURE  MAIN;
   Declarations of all major variables.
   BEGIN (* MAIN *)
   A( ... );
   B( ... );
   END; (* MAIN *)

BEGIN (* DISINHERIT *)
MAIN
END. (* DISINHERIT *)
```

The program merely calls the MAIN procedure. This procedure contains what would normally have been the body of the program, but now there are no global variables; procedures can only communicate information purposely, by means of parameters. Incidentally, the program may also be easier to read because the variable declarations are not separated (by procedure definitions) from the statements that use them.

The second information-hiding technique is as straightforward as the first: always use procedures and functions to access the major data structures. For example, use PUSH and POP procedures to manipulate a stack; do not refer directly to the components of the stack within the program (only in the procedures). In this way there will be no unexpected changes to the data structures, and these structures can be replaced or modified without affecting the overall behavior of the procedures or the integrity of the program. Note that the data structures must be declared as globals or in the MAIN procedure described above, so that their values will be retained between procedure calls.

This leads to another method. Assuming that you really want to use some global variables for the reason just mentioned, or because they are to be used by all or most of the procedures, they can be protected by declaring them within a record this way:

```
VAR
    GLOBAL :
        RECORD
        V1 : ...;
        V2 : ...
        END;
```

The procedures that use V1 and V2 must refer to them as GLOBAL.V1 and GLOBAL.V2, thus protecting them from confusion with local variables—especially locals whose declarations are inadvertently missing. The record notation also clearly identifies globals for the reader.

Yet another contribution to the goals of information hiding can be had by wrapping up any *related variables* into a record. If we recognize, for example, that a list is an array of values *and* an associated variable telling how many elements are in use, we could define:

```
TYPE
    LIST =
        RECORD
        VAL : ARRAY[1..MAX] OF INTEGER;
        SIZE : 0..MAX
        END;
```

Now every variable and parameter of type LIST carries its own SIZE with it. This reduces the number of parameters needed when calling a procedure, protects the LIST components the way globals were protected, and allows the use of the most mnemonic or sensible name (like SIZE) as a component of *all* LIST objects. It also reduces the proliferation of similar names such as ASIZE, BSIZE, CSIZE, and so on. All the data structures used in the algorithms of Part VII are declared that way.

The last information-hiding strategy is more complicated than any of the others. (It is adapted from an article by Michael Feldman; see the Bibliography.) Suppose that we wish to define a data structure and operations for a COMPLEX type. We can do so this way:

```
TYPE
   COMPLEX = @COMPREC;
   COMPREC =
      RECORD
      RE : REAL;            (* real component                    *)
      IM : REAL             (* imaginary component               *)
      END;

VAR
   A, B, C : COMPLEX;

FUNCTION  CADD( A, B : COMPLEX ): COMPLEX;
   (* Performs complex addition of A and B.                      *)

   VAR
      TEMP : COMPLEX;     (* Temporary for result.               *)

   BEGIN (* CADD *)
   (* The function allocates and returns a pointer to a          *)
   (* record containing the result of the addition.              *)
   NEW( TEMP );
   TEMP@.RE := A@.RE + B@.RE;
   TEMP@.IM := A@.IM + B@.IM;
   CADD := TEMP
   END; (* CADD *)
```

Since COMPLEX variables are really pointers to records instead of the records themselves, it is possible to use the more natural functional notation:

```
C := CADD( A, B )
```

instead of a less direct procedure reference that would otherwise have been necessary:

```
CADD( A, B, C )
```

The functional notation can also be nested in the usual way:

```
C := CADD( CADD(A,B), CADD(A,C) )
```

whereas a procedure notation would require the explicit use of temporary variables to hold intermediate results:

```
CADD( A, B, HOLD1 );
CADD( A, C, HOLD2 );
CADD( HOLD1, HOLD2, C )
```

(Actually, there are implicit temporaries; these and other issues are well explained in the article referenced earlier.) The readability of the notation used within the program, and the protection of the COMPLEX type's details from inadvertent manipulation, make this technique attractive in some situations. (This is a third use of pointers that was not discussed in Part VII.)

All these information-hiding strategies require some extra effort; certainly they depend upon the volition of the programmer. The discipline pays off, however, in the reduction of errors and the creation of programs that are easier to read and to modify.

FORMATTED INPUT

It would sometimes be convenient if we could read whole strings from a text file without using loops, or inspect specific columns of an input line, or retain the trailing blanks on each line (if the system usually removes them). These things can, in fact, be done if we are satisfied with reading only character data.

This "formatted input" strategy can be illustrated with these example declarations:

```
TYPE
   CARDIMAGE =           (* Fixed fields of data whose      *)
      RECORD             (* lengths add up to 80.           *)
      A : PACKED ARRAY[1..10] OF CHAR;
      B : PACKED ARRAY[1..20] OF CHAR;
      C : PACKED ARRAY[1..10] OF CHAR;
      ...and so on, dividing the input image into fields.
      END;

VAR
   CARD : CARDIMAGE;
   CARDFILE : FILE OF CARDIMAGE;
```

Instead of using the standard file INPUT to read the data in question, use the file CARDFILE. If the data are arranged into the fields specified by the declarations, then whole data items (strings) are read without loops. We can also look ahead at any part of the *next* card image by using the file pointer ("buffer variable"). In addition, the trailing blanks on each line are always retained, so any field can easily be tested for missing values.

Of course, the card image could also be defined as simply one array of characters, all of which can be read without a loop.

The drawback to this strategy is that any numeric data must be read as strings and then converted to actual numbers. On the other hand, this allows the numeric data to be read regardless of how erroneous it is. It can then be checked for correctness before it is converted to its numeric value, as shown in Part V.

EXACT-PRECISION FRACTIONS

In some applications it is necessary or desirable to use numbers with exact-precision fractions rather than use the inherently approximate reals. The most common applications are programs that deal with dollars and cents.

The simplest situation to handle is when monetary values are merely added or subtracted. Then we can use declarations like these:

```
TYPE
   FIXED2 = INTEGER;      (* Values with 2 decimal places.  *)

VAR
   COST : FIXED2;
   FEE : FIXED2;
   TOTAL : FIXED2;
```

We simply store the COST, for example, as the actual value times 100, so that it is represented as an integer. Although we usually have to account for the fact that the decimal place is implied, simple addition is straightforward:

```
TOTAL := COST + FEE
```

Output can be accomplished with a procedure that puts the decimal point in the implied position:

```
PROCEDURE  WRITEFIXED( VAL : FIXED2; COLS : INTEGER );
   (* Print the given VAL in COLS columns, as ddd.cc       *)

   VAR
      DOLLARS : INTEGER;   (* The whole part of value VAL. *)
      CENTS : INTEGER;     (* "Fractional" part of VAL.    *)

   BEGIN (* WRITEFIXED *)

   (* Break the value into its parts:                      *)
   DOLLARS := VAL DIV 100;
   CENTS := VAL MOD 100;

   (* Print the two parts with the required decimal point:*)
   WRITE( DOLLARS:(COLS-3), '.' );
   IF CENTS < 10 THEN
      WRITE( '0' );
   WRITE( CENTS:1 )          (* Uses 2 columns if needed.   *)

   END; (* WRITEFIXED *)
```

Input is a little more complicated. Assume that we wish to enter values as dollars only, or as dollars and cents:

```
PROCEDURE  READFIXED( VAR VAL : FIXED2 );
   (* Input a value of the form ddd or ddd.cc, and scale  *)
   (* it to fit the FIXED2 type:                           *)

   VAR
      DOLLARS : INTEGER;  (* The whole part of the input. *)
      CENTS : INTEGER;    (* The fractional part, if any. *)
      SIGN : -1..1;       (* The sign of the input value. *)

   BEGIN (* READFIXED *)

   (* Skip leading blanks, if any:                         *)
   WHILE INPUT@ = ' ' DO
      GET( INPUT );

   (* Read the sign, if any:                               *)
   SIGN := 1;
   IF INPUT@ IN ['-','+'] THEN
      BEGIN (* get sign *)
      IF INPUT@ = '-' THEN
         SIGN := -1;
      GET( INPUT )
      END; (* get sign *)

   (* Get the whole (dollars) part:                        *)
   DOLLARS := 0;
   WHILE INPUT@ IN ['0'..'9'] DO
      BEGIN (* each digit *)
      DOLLARS := DOLLARS*10 + ORD(INPUT@)-ORD('0');
      GET( INPUT )
      END; (* each digit *)

   (* The next input char is a '.' if there are cents:     *)
   CENTS := 0;
   IF INPUT@ = '.' THEN
      BEGIN (* get cents *)
      GET( INPUT );
      READ( CENTS )
      END; (* get cents *)

   (* Scale the dollars:                                   *)
   DOLLARS := DOLLARS * 100;

   (* Assemble the final value, accounting for negatives: *)
   VAL := SIGN * (DOLLARS+CENTS)

   END; (* READFIXED *)
```

Ex. 1: Write a procedure that converts a character-string representation of a number (in the form ddd.cc, perhaps with surrounding blanks) into a value of FIXED2 type.

The capacity to represent exact-precision fractions can be expanded and generalized to larger whole numbers and fractions using a data structure like this:

```
TYPE
   FIXED =
      RECORD                (* Exact precision data type:     *)
      WHOLE : INTEGER;      (* The whole part of the value.   *)
      FRAC : INTEGER;       (* The "fractional" part.         *)
      SCALE : INTEGER       (* The fraction is represented by *)
      END;                  (* the value FRAC/SCALE.          *)
```

The SCALE field is not necessary if every value of FIXED type will be scaled by the same amount, in which case the maximum size of FRAC must be implicit in the program. In this example, the presence of SCALE allows two variables that are scaled differently to be added (for example) and assigned to a third variable having a possibly different scale factor.

A variable of the FIXED type might be initialized to zero this way:

```
PROCEDURE INITFIXED( VAR NUMBER : FIXED; MAG : INTEGER );
   (* Initialize a FIXED number to zero, and establish   *)
   (* the number of decimal places it will have:         *)

   BEGIN (* INITFIXED *)

   WITH NUMBER DO
      BEGIN
      WHOLE := 0;
      FRAC := 0;
      SCALE := MAG
      END

   END; (* INITFIXED *)
```

†*Ex. 2: Write an ADDFIXED procedure that uses the FIXED type but assumes that the parameters all have the same number of decimal places.*

Ex. 3: Rewrite ADDFIXED to allow different scale factors among its parameters.

A complete set of FIXED functions could be implemented, for example, using the last of the information-hiding strategies discussed in a previous section.

RANDOM NUMBER SEQUENCES

Many game-playing programs and simulations require the use of "random numbers" to simulate chance occurrences or random events. These are actually numbers taken from a random *sequence* of numbers produced by a random number generator. There are several kinds of random sequences; they differ with respect to the distributions of their values.

Pascal does not provide random number generators. Therefore, we present several of them in this section. They are merely listed here for convenience.

The basic function RANDOM depends upon a global "random seed" variable that retains a value of the last number in the random sequence. It should be initialized in the program, then left alone. The basic declarations that we need are:

```
CONST
   STARTSEED = 49631;       (* Starting value for random seed.*)
                            (* (Use 0 <= STARTSEED <= 65535)   *)

VAR
   SEED : INTEGER;          (* A seed for the random sequence.*)
```

Since the sequence is generated by an algorithm, it can not be truly random, but because it *looks* random, it is called a "pseudorandom sequence." The function is:

```
FUNCTION  RANDOM( VAR SEED : INTEGER ): REAL;
   (* Returns the next number in a uniformly distributed  *)
   (* random sequence such that 0.0 <= RANDOM < 1.0.       *)
   (* *** SEED is also updated to prepare for the next    *)
   (* value in the sequence.                               *)
   (*                                                      *)
   (* Note that all the other generators call this one.   *)

   BEGIN (* RANDOM *)

   SEED := (13849 + 25173*SEED) MOD 65536;
   RANDOM := SEED / 65536

   END; (* RANDOM *)
```

(The reasons for using the particular constants shown in the function are beyond the scope of this book.) The program initializes the seed once, then uses any of the functions:

```
(* Initialize the random seed:                              *)
SEED := STARTSEED;
 ...
(* Print five numbers in the random sequence:               *)
FOR I:=1 TO 5 DO
   WRITELN( RANDOM(SEED) )
```

Here are the other commonly used random sequence generators. The comments in these functions describe the sequence that each produces:

```
FUNCTION  SELECT( LOW : INTEGER;
                  HIGH : INTEGER;
                  VAR SEED : INTEGER ): INTEGER;
   (* Return an integer selected at random from the range *)
   (* LOW..HIGH:                                          *)

   BEGIN (* SELECT *)

   SELECT := TRUNC( RANDOM(SEED)*(HIGH-LOW) ) + LOW

   END; (* SELECT *)


FUNCTION  NORMAL( VAR SEED : INTEGER; MEAN : REAL;
                  STDEV : REAL ): REAL;
   (* Returns a real number from a sequence normally      *)
   (*distributed around the given MEAN, and having the    *)
   (* given standard deviation STDEV:                     *)

   VAR
      SUM : REAL;
      I : 1..12;

   BEGIN (* NORMAL *)

   SUM := 0.0;
   FOR I:=1 TO 12 DO
      SUM := SUM + RANDOM(SEED);
   NORMAL := STDEV*(SUM-6.0) + MEAN

   END; (* NORMAL *)


FUNCTION  EXPORAND( VAR SEED : INTEGER; MEAN : REAL ): REAL;
   (* Returns a number from an exponentially distributed  *)
   (* sequence with a given mean:                         *)

   BEGIN (* EXPORAND *)

   EXPORAND := -MEAN * LN( RANDOM(SEED) )

   END; (* EXPORAND *)
```

```
FUNCTION  POISSON( VAR SEED : INTEGER;
                   MEAN : REAL ): INTEGER;
   (* Returns a number from a sequence with a Poisson     *)
   (* distribution around a given mean:                   *)

   VAR
      NUM : INTEGER;
      PROD : REAL;
      BASE : REAL;

   BEGIN (* POISSON *)

   NUM := 0;
   BASE := EXP( -MEAN );
   PROD := RANDOM( SEED );
   WHILE PROD >= BASE DO
      BEGIN
      PROD := PROD * RANDOM( SEED );
      NUM := NUM + 1
      END;
   POISSON := NUM

   END; (* POISSON *)
```

Answers to Exercises

ANSWERS FOR PART I

Ex. 2. Just one.

Ex. 5. It would be tedious and error-prone to have to count large amounts of data.

Ex. 7. Assume that NEXT is INTEGER. This algorithm works if there are at least two values in the input:

```
(* Get and echo data up to the pair of sentinels:          *)
READLN( DATUM );
READLN( NEXT );
WHILE NOT ((DATUM = SENTINEL) AND (NEXT = SENTINEL)) DO
   BEGIN (* each value *)
   WRITELN( DATUM );
   (* Shift an input value into the DATUM,NEXT pair:       *)
   DATUM := NEXT;
   READLN( NEXT )
   END (* each value *)
```

Ex. 9. If your Pascal places the EOLN marker right after the last nonblank character on each line, then:

```
(* Get and echo data values until end-of-file:             *)
WHILE NOT EOF(INPUT) DO
   BEGIN (* each value *)
   READ( DATUM );
   WRITE( DATUM );
   (* See if this was the last on the line:                *)
   IF EOLN(INPUT) THEN
      BEGIN (* terminate lines *)
      READLN;
      WRITELN
      END (* terminate lines *)
   END (* each value *)
```

(The Pascal used in this book removes the extra blanks.) If your Pascal keeps all the blanks at the ends of the lines:

```
(* Get and echo data values until end-of-file:              *)
WHILE NOT EOF(INPUT) DO
   BEGIN (* each value *)
   READ( DATUM );
   WRITE( DATUM );
   (* Skip blanks up to the next datum (if any) or EOLN:   *)
   WHILE (INPUT@ = ' ') AND NOT EOLN(INPUT) DO
      GET( INPUT );
   (* See if this is the end-of-line:                       *)
   IF EOLN(INPUT) THEN
      BEGIN (* terminate lines *)
      READLN;
      WRITELN
      END (* terminate lines *)
   END (* each value *)
```

Obviously, it is best to arrange the input so that there is a fixed number of values per line.

Ex. 11. No read is necessary before the EOF loop:

```
(* Get and echo groups of data up to end-of-file:           *)
WHILE NOT EOF(INPUT) DO
   BEGIN (* each group *)
   (* Get and echo values until the end-of-group:           *)
   READ( DATUM );
   WHILE DATUM <> GROUPEND DO
      BEGIN (* each value *)
      WRITE( DATUM );
      READ( DATAUM )
      END; (* each value *)
   WRITELN;
   READLN
   END (* each group *)
```

Ex. 12. This algorithm distinguishes among small, medium, and large values, as an example:

```
(* Get and echo values to one of three output columns      *)
(* until EOF:                                              *)

(* Print column headings:                                  *)
WRITELN( 'Small':10, 'Medium':10, 'Large':10 );
(* Get and echo until end-of-file:                         *)
WHILE NOT EOF(INPUT) DO
   BEGIN (* each value *)
   READLN( DATUM );
   (* Select an output column based upon value ranges:    *)
   IF DATUM >= LARGE THEN
      WRITELN( ' ':10, ' ':10, DATUM:10 )
   ELSE IF DATUM >= MEDIUM THEN
      WRITELN( ' ':10, DATUM:10 )
   ELSE
      WRITELN( DATUM:10 )
   END (* each value *)
```

Ex. 15. Assume two INTEGER counters called GROUPCOUNT and VALUECOUNT:

```
(* Get and echo groups of data up to the final sentinel,  *)
(* and print the number of groups:                         *)
GROUPCOUNT := 0;
READ( DATUM );
WHILE DATUM <> SENTINEL DO
   BEGIN (* each group *)

   (* Count this group:                                    *)
   GROUPCOUNT := GROUPCOUNT + 1;

   (* Get and echo group values until the end of group,   *)
   (* and print the number of values in the group:        *)
   VALUECOUNT := 0;
   WHILE DATUM <> GROUPEND DO
      BEGIN (* each value *)
      (* Count this value, then echo, etc.:               *)
      VALUECOUNT := VALUECOUNT + 1;
      WRITE( DATUM );
      READ( DATUM )
      END; (* each value *)
   WRITELN( '<', VALUECOUNT:2, ' values>' );

   READ( DATUM )
   END; (* each group *)
WRITELN( '***', GROUPCOUNT:2, ' groups***' )
```

Ex. 17. Just read the first datum (if there is one) straight into SMALL, then start the loop.

ANSWERS FOR PART II

Ex. 3. Let I be the resulting INTEGER, and CH be an input CHAR:

```
(* Read a digit sequence and convert it to an integer:    *)
I := 0;
READ( CH );
WHILE CH IN ['0'..'9'] DO
   BEGIN (* each digit *)
   I := I*10 + (ORD(CH)-ORD('0'));
   READ( CH )
   END (* each digit *)
```

(The way Pascal actually reads an integer is similar to this:

```
(* Read a digit sequence and convert it to an integer:    *)
I := 0;
WHILE INPUT@ IN ['0'..'9'] DO
   BEGIN (* each digit *)
   I := I*10 + (ORD(INPUT@)-ORD('0'));
   GET( INPUT )
   END (* each digit *)
```

which does not read the character after the digits.)

Ex. 4. There is a READ before the loop because the inner WHILEs depend upon having the first character ready to test (like any sentinel loop). There is an extra EOF test inside the main loop because the blank(s) being skipped may follow the last word of the last line of input.

Ex. 7. Initialize INAWORD to FALSE (meaning "in blanks") instead of TRUE.

Ex. 8. The declaration depends upon your intentions, but this is an example:

```
TYPE
   CHARCLASSES =
      (BLANK,            (* Just the ' '.                  *)
       ALPHA,            (* 'A'..'Z', and 'a'..'z'         *)
       DIGIT,            (* '0'..'9'                       *)
       OPERATOR,         (* Any of '+-/*=><@'              *)
       DELIMITER,        (* Any of '.,;:(){}[]'            *)
       OTHER);           (* All others                     *)
```

ANSWERS FOR PART III

Ex. 3.

```
(* Read values into list A up to MAX or EOF:                  *)
N := 0;
WHILE (N < MAX) AND NOT EOF(INPUT) DO
   BEGIN (* each value *)
   (* Count the new element and get it directly:              *)
   N := N + 1;
   READLN( A[N] )
   END; (* each value *)

(* You could now check for excess data, if you wish:          *)
IF NOT EOF(INPUT) THEN
   WRITELN( 'WARNING--excess data given as input.' )
```

Ex. 6.

```
(* Print the values in A, their running total, and the       *)
(* values increased by 4%:                                    *)
WRITELN( 'Value':10, 'Running Total':15, 'Value+4%':10 );
SUM := 0;
FOR I:=1 TO N DO
   BEGIN (* each value *)
   SUM := SUM + A[I];
   WRITELN( A[I]:10, SUM:15, (A[I]*1.04):10:2 )
   END (* each value *)
```

Ex. 9. If FOUND is FALSE, POSN is one greater than N.

Ex. 10. POSN begins at N and decreases by one each time through the loop. The looping condition should test for POSN greater than or equal to 1. So, if FOUND remains FALSE, POSN is 0; otherwise POSN points to the value.

Ex. 11.

```
IF FOUND THEN
   IF A[POSN] = X THEN
      (* actually found X *)
   ELSE
      (* just found the place for X *)
ELSE
   (* X would be beyond A[N] *)
```

Ex. 13. This one works if N is 0, but not if N is equal to MAX, since A[N + 1] would not exist. The previous one works in all situations.

Ex. 14. Yes.

Ex. 15. If X is larger than all values, LOW is one greater than N; if X is smaller than all values, LOW is 1.

Ex. 19. The value at position POSN would be duplicated throughout that group of items in A.

Ex. 21. It works when POSN is one greater than N, even if N is 0.

Ex. 25. Here the FOR loop is not executed if POSN is greater than N, although the assignment of X is redundant in that case:

```
(* Append or insert X, given the result of the search:     *)
FOR I:=N DOWNTO POSN DO
   A[I+1] := A[I];
N := N+1;
A[POSN] := X
```

Ex. 28. Because A[J + 1] = A[(I-1) + 1] = A[I] is where the item came from, and A[POSN + 1] is where the "gap" should be left to receive it.

Ex. 31.

```
(* Delete all items whose values are less than LIMIT:      *)

(* Find the position of the smallest value to be kept:     *)
POSN := 1;
FOUND := FALSE;
WHILE (POSN <= N) AND NOT FOUND DO
   IF A[POSN] >= LIMIT THEN
      FOUND := TRUE
   ELSE
      POSN := POSN + 1;

(* Shift the good values all the way to the front of A:    *)
FOR I:=POSN TO N DO
   A[I-POSN] := A[I];

(* Reset N to the number of items retained:                *)
N := N - POSN + 1
```

```
(* Delete all items whose values are larger than LIMIT:    *)

(* Search backward for position of largest value to keep: *)
POSN := N;
FOUND := FALSE;
WHILE (POSN >= 1) AND NOT FOUND DO
   IF A[POSN] <= LIMIT THEN
      FOUND := TRUE
   ELSE
      POSN := POSN - 1;

(* Reset N to the number of items kept:                    *)
N := POSN
```

Ex. 33. Change the given merge this way: initialize K to 0, then replace each WRITELN with, for example:

```
(* WRITELN( A[APOSN] ) becomes:                            *)
K := K + 1;
C[K] := A[APOSN]
```

Ex. 35. Search for the breakoff point in A, and at the same time copy the values into B. Then shift the remaining elements of A down to the first position.

Ex. 38. If their lengths are unequal, the lists are not equal. To see if corresponding elements match, use the length of the shorter list to limit the loop.

Ex. 40. For example:

```
(* Pop the top of STK into X, if possible:                 *)
WITH STK DO
   IF TOP > 0 THEN
      BEGIN (* pop *)
      X := VAL[TOP];
      TOP := TOP - 1
      END (* pop *)
   ELSE
      WRITELN( 'ERROR--attempt to pop an empty stack.' )
```

ANSWERS FOR PART IV

Ex. 1. It allows the input to have more than NCOLS values per line, so that different amounts of data can be used in testing--without preparing new data each time.

Ex. 4. The variable SUM can be used to hold the sum of each row--if each row and its sum are printed immediately.

Ex. 5. (b) The sentinels can be kept in each row; the lengths of all rows can be stored in an extra column set aside for them; the lengths could be saved in an extra array (of type ONECOLUMN).

Ex. 6. The column sums must all be initialized at the beginning:

```
(* Find and store the column sums of table M:                    *)
FOR COL:=1 TO NCOLS DO
   COLSUM[COL] := 0;
FOR ROW:=1 TO NROWS DO
   FOR COL:=1 TO NCOLS DO
      COLSUM[COL] := COLSUM[COL] + M[ROW,COL]
```

Ex. 9. To exchange consecutive columns, for example, use the variable SAVE of type ITEM:

```
(* Exchange columns COL and COL+1 of M:                          *)
FOR ROW:=1 TO NROWS DO
   BEGIN (* swap a pair *)
   SAVE := M[ROW,COL];
   M[ROW,COL] := M[ROW,COL+1];
   M[ROW,COL+1] := SAVE
   END (* swap a pair *)
```

Ex. 13. The search would look at the last value in each row to find the proper row, then look within that row to find the column position.

Ex. 16. Using an insertion sort:

```
(* Sort the table T into ascending order by keys:          *)
FOR ROW:=2 TO NROWS DO
   BEGIN (* each row *)
   HOLD := T[ROW];
   POSN := ROW - 1;
   FOUND := FALSE;
   WHILE (POSN >=1) AND NOT FOUND DO
      BEGIN (* search and shift *)
      IF HOLD.KEY < T[POSN].KEY THEN
         BEGIN (* shift *)
         T[POSN+1] := T[POSN];
         POSN := POSN - 1
         END
      ELSE
         FOUND := TRUE
      END; (* search and shift *)
   T[POSN+1] := HOLD
   END; (* each row *)
```

ANSWERS FOR PART V

Ex. 2. This procedure also skips over any spaces before the quoted string, which allows a string, like an integer, to occur anywhere on the input line:

```
PROCEDURE  READQUOTED( VAR S : STRING );
   (* Read a quoted string (without embedded quotes).       *)
   (* Preceding spaces are skipped.                         *)

   CONST
      QUOTE = '''';

   VAR
      CH : CHAR;           (* One character of input.      *)

   BEGIN (* READQUOTED *)

   (* Use the first quote as a sentinel for the leading    *)
   (* spaces (if any):                                     *)
   READ( CH );
   WHILE CH <> QUOTE DO
      READ( CH );
   (* Use the closing quote as a sentinel for the string: *)
   S := NULL;
   READ( CH );
   WITH S DO
      WHILE CH <> QUOTE DO
         BEGIN (* each char *)
         LEN := LEN + 1;
         STR[LEN] := CH;
         READ( CH )
         END (* each char *)

   END; (* READQUOTED *)
```

Ex. 4. The function (call it VERIFY, for example) is useful for checking strings to see if they contain only the proper characters. For example, the string 'Zapata, E.' passes the test:

```
IF VERIFY( S, ['A'..'Z','a'..'z',' ',',','.'] ) = 0 THEN
   ...
```

whereas the string 'Br0wn, 1.' would not. The function would return a 3.

Ex. 9.

```
FUNCTION  COMPARE( A : STRING; B :STRING ) : INTEGER;
   (* Return -1 if A<B, 0 if A=B, or 1 if A>B, where       *)
   (* 'x' < 'x ', etc.                                     *)

   BEGIN (* COMPARE *)

   IF A.STR < B.STR THEN
      COMPARE := -1
   ELSE IF A.STR > B.STR THEN
      COMPARE := 1
   ELSE
      IF A.LEN < B.LEN THEN
         COMPARE := -1
      ELSE IF A.LEN > B.LEN THEN
         COMPARE := 1
      ELSE
         COMPARE := 0

   END; (* COMPARE *)
```

Ex. 11. Two advantages: the input loop is simplified and protected from excess data. A disadvantage is that the total program is longer than necessary.

Ex. 14. The simplest form (without error checking) would be:

```
PROCEDURE  INSERT( NEW : STRING; VAR OLD : STRING;
                                     POS : INTEGER );
   (* Insert a copy of NEW into OLD at position POS, thus *)
   (* making OLD longer.                                  *)

   VAR
      PART1 : STRING;    (* Substring of OLD prior to POS. *)
      PART2 : STRING;    (* Substring from POS to the end. *)

   BEGIN (* INSERT *)

   COPY( OLD, 1, POS-1, PART1 );
   COPY( OLD, POS, (OLD.LEN-POS+1), PART2 );
   OLD := PART1;
   CATSTRING( OLD, NEW );
   CATSTRING( OLD, PART2 )

   END; (* INSERT *)
```

Ex. 15. Yes.

Ex. 17.

```
PROCEDURE  MAKESTRING( VAR S : STRING;
                       SOURCE : FIXEDSTRING );
   (* Put a trimmed varying string representation of      *)
   (* SOURCE into S.                                      *)

   BEGIN (* MAKESTRING *)

   S.STR := SOURCE;
   S.LEN := MAXLEN;
   TRIM( S )

   END; (* MAKESTRING *)
```

Ex. 24. The function would treat the digits up to the decimal point as an integer. The fractional part is done separately, then added to the whole part to give the total real value. It would be most accurate to isolate the fractional part as an integer, then divide it by the appropriate power of 10.

ANSWERS FOR PART VI

Ex. 2. Either there would have to be another parameter (BOOLEAN) to tell the calling program whether or not the key was found, *or* a special value could be returned in I.INFO to signal failure. In either case, the caller could not always use I.INFO as good data. Assuming an extra parameter called FOUND, the search would become:

```
FOUND := FALSE;
WHILE NOT EOF(F) AND NOT FOUND DO
   IF F@.KEY <> I.KEY THEN
      GET( F )
   ELSE
      FOUND := TRUE
```

Ex. 5. No temporary file is necessary here:

```
PROCEDURE  SPLIT( VAR OLD : ITEMFILE; VAR NEW : ITEMFILE );
   (* Copy to file NEW all items of OLD whose keys are in *)
   (* the range LOW..HIGH (global constants).             *)

   BEGIN (* SPLIT *)

   (* Prepare the files for I/O:                          *)
   RESET( OLD );
   REWRITE( NEW );
   (* Look through all of the keys in OLD until EOF:      *)
   WHILE NOT EOF(OLD) DO
      BEGIN (* each item *)
      (* Copy the item to NEW if necessary:               *)
      IF OLD@.KEY IN [LOW..HIGH] THEN
         BEGIN (* copy *)
         NEW@ := OLD@;
         PUT( NEW )
         END; (* copy *)
      (* Step to the next item in OLD, regardless:        *)
      GET( OLD )
      END (* each item *)

   END; (* split *)
```

Ex. 6. This will require a temporary file--and other changes.

Ex. 8. This is like INSERT except that when the given key is found, the item is *not* copied to the TEMP file. (The condition in the IF statement must test for equality.)

Ex. 10. (b)

```
(* Initialize LF from data given as standard input (one    *)
(* input item per line--assuming scalars):                  *)
REWRITE( LF );
ILIST.COUNT := 0;
WHILE NOT EOF(INPUT) DO
   BEGIN (* each item *)
   (* Get the item from input:                              *)
   WITH I DO
      READLN( KEY, INFO );
   (* Accumulate the item into the current list, if         *)
   (* possible.  Otherwise, start a new list:               *)
   WITH ILIST DO
      IF COUNT < 50 THEN
         BEGIN (* accumulate *)
         COUNT := COUNT + 1;
         LIST[COUNT] := I
         END (* accumulate *)
      ELSE
         BEGIN (* start new list *)
         WRITE( IF, ILIST );
         COUNT := 1;
         LIST[COUNT] := I
         END (* start new list *)
   END; (* each item *)
(* Output the remaining (usually partly full) list:         *)
WRITE( LF, ILIST )
```

Ex. 11. Insert an additional test before the ELSE alternative already there:

```
ELSE IF MASTER@.KEY = NEW@.KEY THEN
   BEGIN (* duplicate keys *)
   (* Take item from MASTER, but step along both files:   *)
   TEMP@ := MASTER@;
   PUT( TEMP );
   GET( MASTER );
   GET( NEW )
   END (* duplicate keys *)
```

Ex. 12. The reason is that delete transactions are supposed to have keys that match master file keys. Warning messages should be printed for leftover deletes, but they should otherwise be ignored. (By the definition of insertions, there is no such thing as leftover change transactions.)

Ex. 16. Although the algorithm would not be very efficient, it uses the concept of runs. Here is an outline:

```
(* Merge the unordered file NEW into the ordered file      *)
(* OLD, maintaining the order:                             *)
RESET( NEW );
WHILE NOT EOF(NEW) DO
   BEGIN
   RESET( OLD );
   REWRITE( TEMP );
   Merge OLD and the next run from NEW into TEMP;
   COPYFILE( TEMP, OLD )
   END
```

Ex. 18. Assume that SUM is type INFOTYPE. Most of the body of the procedure would be changed:

```
...
IF ANY THEN
   BEGIN (* get first *)
   READ( F, HOLD );
   SUM := HOLD.INFO
   END; (* get first *)

(* Step through the items, writing them only when the   *)
(* key changes, otherwise summing the INFO fields:      *)
WHILE NOT EOF(F) DO
   (* Compare the next key to the one held back:        *)
   IF F@.KEY <> HOLD.KEY THEN
      BEGIN (* key change *)
      (* Write the one held back (putting SUM as its    *)
      (* INFO field), then get and hold the next:       *)
      HOLD.INFO := SUM;
      WRITE( TEMP, HOLD );
      READ( F, HOLD );
      SUM := HOLD.INFO
      END (* key change *)
   ELSE
      BEGIN (* same key *)
      (* Add its INFO to the sum, then skip it:         *)
      SUM := SUM + F@.INFO;
      GET( F )
      END; (* same key *)

(* If there were any at all, output the one held back: *)
IF ANY THEN
   BEGIN (* last one *)
   HOLD.INFO := SUM;
   WRITE( TEMP, HOLD )
   END; (* last one *)
...
```

ANSWERS FOR PART VII

Ex. 2. POP (for example) can be rewritten so that if the stack is empty, a special value is returned through ITEM, or a third parameter (BOOLEAN) could be used to tell the caller whether the pop was successful. In any case, the ITEM returned may not be usable data, so the caller must check to see if it is. It is conceptually better just to use the STACKEMPTY function to see if POP should be called at all.

Ex. 3. Just drop the FRONT pointer from the declarations. Here is an example procedure:

```
PROCEDURE  REMOVE( VAR Q : QUEUE; VAR ITEM : INFOTYPE );
   (* Remove the node from the front of the circular      *)
   (* queue, and return its value in ITEM.                *)

   VAR
      FRONT : NODEPTR;   (* Temporary pointer to the front.*)

   BEGIN (* REMOVE *)

   (* The pointer within the rear node points to the      *)
   (* front node.  Locate the front and copy the value:   *)
   FRONT := Q.REAR@.NEXT;
   ITEM := FRONT@.INFO;
   (* Remove the front node from the list:                *)
   Q.REAR@.NEXT := FRONT@.NEXT;
   (* Get rid of the old front node:                      *)
   IF Q.REAR = FRONT THEN
      (* There was only one node (it was both the front   *)
      (* and the rear):                                   *)
      Q.REAR := NIL;
   DISPOSE( FRONT )

   END; (* REMOVE *)
```

Ex. 5. The PRINTLIST procedure (for example) would need two changes to support the new implementation: change the test for HEAD = SENTINEL to HEAD = NIL (the new definition of an empty list), and change UNTIL P = SENTINEL to UNTIL P = NIL (the pointer P stepped off the end of the list).

Ex. 8.

```
(* Build an ordered (ascending) linked list from          *)
(* unordered input (one scalar value per line):           *)
LISTINIT( LIST );
WHILE NOT EOF(INPUT) DO
   BEGIN (* each item *)
   READLN( ITEM );
   P := SEARCH( LIST, ITEM );   (* P : NODEPTR            *)
   INSERT( ITEM, LIST, P )
   END (* each item *)
```

Ex. 9. One way to order the list is to use an algorithm similar to that of Ex. 8 to build a new list, taking the values from the old list instead of from input. Another way is to step through the list looking at pairs of consecutive nodes, swapping their contents into order when necessary (bubble sort).

Ex. 11. If this search algorithm is used:

```
P := HEAD;
FOUND := FALSE;
WHILE (P <> NIL) AND NOT FOUND DO
   BEGIN (* each node *)
   IF P@.INFO >= ITEM THEN
      FOUND := TRUE
   ELSE
      P := P@.NEXT
   END (* each node *)
```

then P will in the end be NIL when the given item is larger than every value in the list. So P can not always be used to show where a new node should be inserted into the list. This problem is solved by having an extra pointer, PREV, that always points to the node previous to the one being compared:

```
P := HEAD;
PREV := HEAD;
FOUND := FALSE;
WHILE (P <> NIL) AND NOT FOUND DO
   BEGIN (* each node *)
   IF P@.INFO >= ITEM THEN
      FOUND := TRUE
   ELSE
      BEGIN (* step along *)
      PREV := P;
      P := P@.NEXT
      END (* step along *)
   END (* each node *)
```

A new node can now very simply be inserted after PREV and before P (the simple exceptions are when PREV = HEAD). No swapping of node values is necessary with this list implementation.

Ex. 13.

```
PROCEDURE  CONDENSE( LIST : LINKLIST;
                     VAR NEWLIST : LINKLIST );
   (* Return in NEWLIST a list of all the different          *)
   (* values from LIST, which is ordered.                    *)

   VAR
      PREV : NODEPTR;     (* Both are for stepping through   *)
      P : NODEPTR;        (* LIST, keeping PREV behind P.    *)

   BEGIN (* CONDENSE *)

   LISTINIT( NEWLIST );
   (* Put the first item of LIST, if any, into NEWLIST:      *)
   P := LIST.HEAD;
   PREV := P;
   IF P <> LIST.SENTINEL THEN
      BEGIN (* save first *)
      INSERT( P@.INFO, NEWLIST, NEWLIST.SENTINEL );
      P := P@.NEXT
      END; (* save first *)
   (* Step through LIST, putting the next value into         *)
   (* NEWLIST if it is not equal to the previous value:      *)
   WHILE P <> LIST.SENTINEL DO
      BEGIN (* each item *)
      IF P@.INFO <> PREV@.INFO THEN
         INSERT( P@.INFO, NEWLIST, NEWLIST.SENTINEL );
      PREV := P;
      P := P@.NEXT
      END (* each item *)

   END; (* CONDENSE *)
```

Ex. 16. The body of the procedure becomes:

```
   ...
   (* Process this tree if it is nonempty:                    *)
   IF P <> NIL THEN
      BEGIN (* nonempty tree *)
      (* Print all values less than this one, if any:         *)
      PRINTNODES( P@.LEFT );
      (* Print this one:                                      *)
      WRITELN( P@.INFO );
      (* Print all values greater than this one, if any:      *)
      PRINTNODES( P@.RIGHT )
      END (* nonempty tree *)
```

ANSWER FOR PART VIII

Ex. 2.

```
PROCEDURE ADDFIXED( X : FIXED; Y : FIXED; VAR SUM : FIXED );
   (* Find the sum of the fixed decimal values X and Y.     *)
   (* X and Y are assumed to be positive.                   *)

   BEGIN (* ADDFIXED *)

   INITFIXED( SUM, X.SCALE );
   (* Find the fractional part and the carry (if any):      *)
   SUM.FRAC := X.FRAC + Y.FRAC;
   SUM.WHOLE := SUM.FRAC DIV SUM.SCALE;
   SUM.FRAC := SUM.FRAC MOD SUM.SCALE;
   (* Add the whole parts and the carry:                    *)
   SUM.WHOLE := SUM.WHOLE + X.WHOLE + Y.WHOLE

   END; (* ADDFIXED *)
```

Bibliography

STANDARD AND UCSD PASCAL REFERENCES

Jensen, K., and N. Wirth, *Pascal User Manual and Report.* Springer-Verlag, 1974.

Addyman, A. M., et al., "A Draft Proposal for Pascal," *SIGPLAN Notices,* Vol. 15, No. 4, pp. 1-66 (1980).

Clark, R., and S. Koehler, *The UCSD Pascal Handbook.* Reward/Reston, 1982.

Cooper, D., *Standard Pascal User Reference Manual.* Norton, 1983.

Tiberghien, J., *The Pascal Handbook.* Sybex, 1981.

INTRODUCTORY PASCAL PROGRAMMING TEXTS (STANDARD AND UCSD PASCAL)

Atkinson, L. V., *Pascal Programming.* Wiley, 1980.

Brainerd, W. S., C. H. Goldberg, and J. L. Gross, *Pascal Programming: A Spiral Approach.* Boyd & Fraser, 1982.

Cherry, G. W., *Pascal Programming Structures: An Introduction to Systematic Programming.* Reston, 1980.

Cooper, D., and M. Clancy, *Oh! Pascal!* Norton, 1982.

Cooper, J. W., *Introduction to Pascal for Scientists.* Wiley, 1981.

Dale, N., and D. Orshalick, *Introduction to Pascal and Structured Design.* Heath, 1983.

Dyck, V. A., J. D. Lawson, J. A. Smith, and R. J. Beach, *Computing: An Introduction to Structured Problem Solving Using Pascal.* Reston, 1982.

Findlay, W., and D. F. Watt, *Pascal: An Introduction to Methodical Programming,* 2d ed. Computer Science Press, 1982.

Forsyth, R. S., *Pascal at Work and Play.* Chapman and Hall, 1982.

Gear, C. W., *Programming in Pascal.* Science Research Associates, 1983.

Glinert, E. P., *Introduction to Computer Science Using Pascal.* Prentice-Hall, 1983.

Graham, N., *Introduction to Pascal.* 2d ed. West, 1983

Grogono, P., *Programming in Pascal,* rev. ed. Addison-Wesley, 1980.

Holt, R. C., and J. N. P. Hume, *Programming Standard Pascal.* Reston, 1980.

Hume, J. N. P., and R. C. Holt, *UCSD Pascal: A Beginner's Guide to Programming Microcomputers.* Reward/Reston, 1982.

Jones, R. M., *Introduction to Pascal and Computer Applications.* Allyn and Bacon, 1983.

Keller, A. M., *A First Course in Computer Programming using Pascal.* McGraw-Hill, 1982.

Kemp, R., *Pascal for Students.* Arnold, 1982.

Kennedy, M., and M. B. Solomon, *Pascal Program Development with Ten Instruction Pascal Subset (TIPS) and Standard Pascal.* Prentice-Hall, 1982.

Koffman, E. B., *Problem Solving and Structured Programming in Pascal.* Addison-Wesley, 1981.

—————— *Pascal: A Problem Solving Approach.* Addison-Wesley, 1982.

Ledgard, H., and A. Singer, *Elementary Pascal.* Systems Research Associates, 1982.

Ledin, G., *Pascal.* Alfred, 1982.

Matuszek, D. L., *Quick Pascal.* Wiley, 1982.

Mazlack, L. J., *Structured Problem Solving with Pascal.* Holt, Rinehart and Winston, 1983.

Mendelson, B., *A First Course in Programming with Pascal.* Allyn and Bacon, 1982.

Moore, J. B., *Pascal: Text and Reference with Waterloo Pascal and Pascal VS.* Reston, 1982.

Moore, L., *Foundations of Programming with Pascal.* Halsted/Wiley, 1980.

Prather, R. E., *Problem-Solving Principles: Programming with Pascal.* Prentice-Hall, 1982.

Richards, J. L., *Pascal.* Academic, 1982.

Rohl, J. S., and H. J. Barrett, *Programming via Pascal.* Cambridge, 1980.

Schneider, G. M., S. W. Weingart and D. M. Perlman, *An Introduction to Programming and Problem Solving With Pascal* 2d ed. Wiley, 1982.

Tucker, A. B., *Introduction to Programming with ESP and Pascal.* Holt, Rinehart and Winston, 1983.

Walker, B. K., *A Structured Approach to Pascal.* Irwin, 1983.

Welsh, J., and J. Elder, *Introduction to Pascal.* 2d ed. Prentice-Hall, 1982.

Wilson, I. R., and A. M. Addyman, *A Practical Introduction to Pascal.* 2d ed. Springer-Verlag, 1982.

Zaks, R., *An Introduction to Pascal (Including UCSD Pascal).* Sybex, 1980.

ALGORITHMS AND PROGRAMS IN PASCAL

Aho, A. V., J. E. Hopcroft and J. D. Ullman, *Data Structures and Algorithms.* Addison-Wesley, 1983.

Alagic, S., and M. A. Arbib, *The Design of Well Structured and Correct Programs.* Springer-Verlag, 1978.

Barnard, D. T., and R. G. Crawford, *Pascal Programming Problems and Applications.* Reston, 1982.

Coleman, D., *A Structured Programming Approach to Data.* Springer-Verlag, 1979.

Davidson, G., *Practical Pascal Programs.* Osborne/McGraw-Hill, 1982.

Dromey, R. G., *How to Solve it by Computer.* Prentice-Hall, 1982.

Feldman, M. B., "Information Hiding in Pascal: Packages and Pointers," *BYTE. Vol. 6,* Vol. 6, No. 11, pp. 493-99, (1981).

Hergert, D., and J. T. Kalash, *Apple Pascal Games.* Sybex, 1981.

Hergert, R., and D. Hergert, *Doing Business with Pascal.* Sybex, 1983.

Jones, W. B., *Programming Concepts: A Second Course.* Prentice-Hall, 1982.

Kernighan, B. W., and P. J. Plauger, *Software Tools in Pascal.* Addison-Wesley, 1981.

Marney-Petix, V. (ed.), *Some Common Pascal Programs.* Osborne/McGraw-Hill, 1982.

Miller, A., *Pascal Programs for Scientists and Engineers.* Sybex, 1981.

Schneider, G. M., and S. C. Bruell, *Advanced Programming and Problem Solving with Pascal.* Wiley, 1981.

Tenenbaum, A. M., and M. J. Augenstein, *Data Structures Using Pascal.* Prentice-Hall, 1981.

Wirth, N., *Algorithms + Data Structures = Programs.* Prentice-Hall, 1976.

———— *Systematic Programming: An Introduction.* Prentice-Hall, 1973.

———— *A Collection of Pascal Programs.* Berichte Nr. 33, ETH, Zurich, 1979.

ALGORITHMS IN GENERAL

Aho, A. V., J. E. Hopcroft, and J. D. Ullman, *The Design and Analysis of Computer Algorithms.* Addison-Wesley, 1974.

Baase, S., *Computer Algorithms: Introduction to Design Analysis.* Addison-Wesley, 1979.

Goodman, S. E., and S. T. Hedetniemi, *Introduction to the Design and Analysis of Algorithms.* McGraw-Hill, 1977.

Horowitz, E., and S. Sahni, *Algorithms: Design and Analysis.* Computer Science Press, 1977.

Knuth, D. E., *The Art of Computer Programming.* Vol. 1, *Fundamental Algorithms* 2d ed.; Vol. 2, *Seminumerical Algorithms* 2d ed.; Vol. 3, *Sorting and Searching.* Addison-Wesley, 1973.

Maurer, H. A., and M. R. Williams, *A Collection of Programming Problems and Techniques.* Prentice-Hall, 1972.

Index